A SKY
WONDERFUL
WITH STARS

For my part I know nothing
with any certainty but the sight
of the stars makes me dream.

—VINCENT VAN GOGH

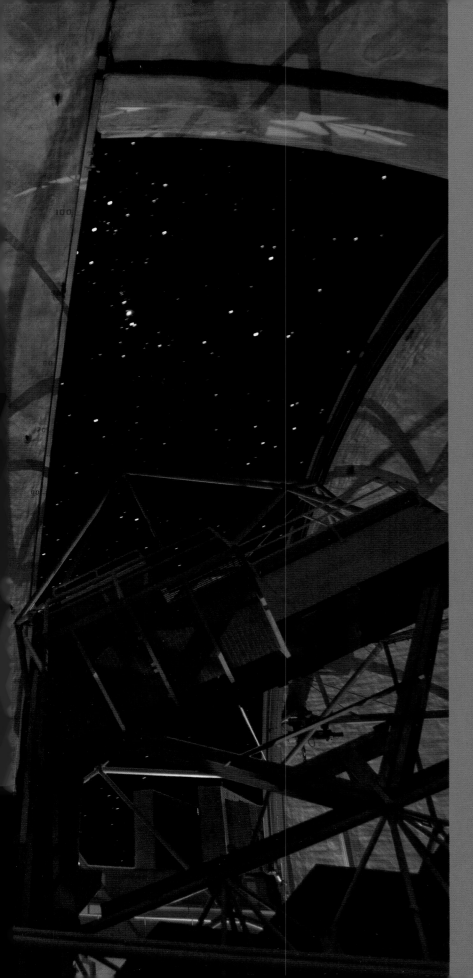

A SKY
WONDERFUL
WITH STARS

50 YEARS *of*
MODERN ASTRONOMY
on MAUNAKEA

MICHAEL J. WEST

A Latitude 20 Book
University of Hawai'i Press
HONOLULU

PHOTO ON TITLE PAGE:

Gateway to the Heavens

PHOTO:
W. M. KECK OBSERVATORY
AND RICK PETERSON

20 19 18 17 16 15   6 5 4 3 2 1

Library of Congress Cataloging-in-Publication Data

West, Michael J., author.
 A sky wonderful with stars : 50 years of modern astronomy on Maunakea / Michael J. West.
    pages cm
 "A latitude 20 book."
 Includes index.
 ISBN 978-0-8248-5268-9 (cloth : alk. paper)
 1. Astronomy—Hawaii—Mauna Kea—History—Pictorial works.   2. Astronomical observatories—
Hawaii—Mauna Kea—History—Pictorial works.   I. Title.
 QB33.U6W47 2015
 522'.199691—dc23

                                                                                2014047491

University of Hawai'i Press books are printed on acid-free paper and meet the guidelines for
permanence and durability of the Council on Library Resources.

Designed by Mardee Melton
Printed by Regent Publishing Services

CONTENTS

FOREWORD

The people of Hawai'i have lent Maunakea to the world, where humanity can look up to the heavens and explore our place in the universe. A desert 4,205 meters atop an island paradise, Maunakea is a unique place on our planet where the land pokes through the smooth winds of the Pacific, allowing a portal through the atmosphere to view the stars with unparalleled detail. It is where astronomy has put its largest telescopes.

Maunakea's skies have led to the discovery of everything from the most distant objects in the universe in the form of gamma ray bursts and quasars to the class of asteroids—Kuiper-belt objects—to which Pluto was demoted from its previous status of planet. Maunakea has allowed us to weigh the black hole in our Milky Way's center and has shown us how galaxies have evolved over the past 13 billion years. Maunakea is where many of the observations were taken that enabled us to measure that the expansion rate of the universe was accelerating.

Michael West captures the drama and beauty of Maunakea as never before in this collection of images and stories. Through them, he tells the story of Maunakea, and through its skies, the history of the universe. Maunakea truly is *a place like no other.*

BRIAN P. SCHMIDT, 2011 Nobel Laureate in Physics

PREFACE

Starlight rains continuously on every square inch of our planet's surface. The ground, the air, and the seas swallow most of it. But we humans manage to collect a tiny fraction of this light with telescopes, and from those precious luminous droplets we learn about the universe and our place in it.

More than sixteen thousand nights have passed since the first telescopes arrived on Maunakea. In that time, the observatories huddled together on the summit of this ancient volcano have been at the forefront of astronomical research. Thousands of scientific papers have been published based on observations made with these telescopes, including the discovery of myriad planets around other stars, the most distant known galaxies, and the accelerating expansion of the universe that led to the 2011 Nobel Prize in Physics. The unprecedented light-gathering power of the giant Thirty Meter Telescope (TMT), to be built in the near future, will allow astronomers to see farther and better than ever before, creating a potent synergy between it and the other telescopes on Maunakea.

But astronomy isn't just the study of distant planets, stars, and galaxies. It's also the study of something much closer to home—us. One of astronomy's most profound discoveries is that we humans are made from the ashes of stars whose fires burned out long ago. We are truly star stuff, "bits of stellar matter that got cold," in the words of twentieth-century astronomer Arthur Eddington. Perhaps that's why we feel compelled to explore the starry skies, as if driven by an innate yearning to know our true ancestral home and ourselves. "You are that vast thing that you see far, far off with the great telescopes," wrote the philosopher Alan Watts.

On the summit of Maunakea, in one of the most unique places on earth, an international village gathers the soft rain of light that falls from the heavens. There have been many challenges in the past, and there will be others in the future as technology, knowledge, and understanding are pushed to new limits. But, as poet Pablo Neruda said, our desire to "play with the light of the universe" is strong. The next fifty years of astronomy on Maunakea will continue a journey begun long ago, an enduring legacy to the human spirit of exploration.

About This Book

An old adage says that every picture tells a story, and this belief has guided the creation of *A Sky Wonderful with Stars*. Because astronomy is the most visual of all sciences, and Hawai'i is one of the most photogenic places on earth, it seemed only natural to tell the story of modern astronomy on Maunakea through photographs accompanied by light text to provide context and information.

The pages that follow are divided into three sections. The first, *A Place Like No Other,* shows why Maunakea is so special. The second, *From Dream to Reality,* shows how this remote mountaintop became home to the most powerful collection of telescopes in the world. And the third, *Postcards from the Universe,* showcases some of the many stunning images and scientific highlights produced by the telescopes on Maunakea.

It's my sincere hope that you'll enjoy this book and that all who choose to visit Maunakea will do so in a way that is respectful of the Hawaiian culture, the natural beauty, and the scientific activities of this unique place.

ACKNOWLEDGMENTS

Many people have contributed in different ways to this book, and to all I am grateful. First and foremost, I thank the Maunakea observatories for their support and their faith in me, in particular Ann Boesgaard, Mike Connelley, Gary Davis, Roy Gal, Debbie Goodwin, John Hamilton, Saeko Hayashi, Stewart Hunter, Steve Jefferson, Tom Kerr, Ka'iu Kimura, Nadine Manset, Pierre Martin, Bob McLaren, Peter Michaud, Harriet Parsons, Glen Petitpas, Tom Phillips, Joy Pollard, Doug Simons, Gordon Squires, John Stoke, Marianne Takamiya, Alan Tokunaga, and Richard Wainscoat. Additionally, I thank everyone who also allowed their spectacular photographs to be included in this book, especially Alexandra Angelich, Jason Chu, Jean-Charles Cuillandre, R. Jay GaBany, Lisa Martin, Sergio Martín Ruiz, Jennifer Miller, Barbara Schaefer, and Peter Tuthill. Thanks, too, to Brian Schmidt for kindly writing the foreword. My heartfelt gratitude goes to Nadine Little, Michael Duckworth, and everyone at University of Hawai'i Press for their enthusiastic support for this project, and to copy editor Lee Motteler and the anonymous reviewers for feedback that helped to improve the book. Thanks also to Govert Schilling for suggesting the idea of a book to commemorate the first fifty years of modern astronomy on Maunakea. Finally, as always, I thank my wife, Cheryl, and my son, Caden, for their patience and, most importantly, for their love.

MAP OF MAUNAKEA OBSERVATORIES

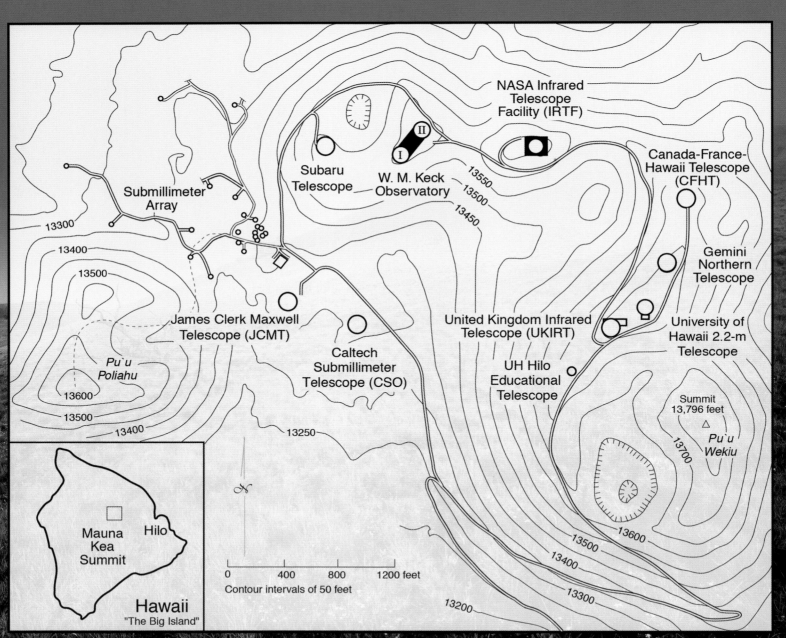

NASA Infrared Telescope Facility (IRTF)

Canada-France-Hawaii Telescope (CFHT)

Subaru Telescope

W. M. Keck Observatory

Submillimeter Array

Gemini Northern Telescope

James Clerk Maxwell Telescope (JCMT)

United Kingdom Infrared Telescope (UKIRT)

University of Hawaii 2.2-m Telescope

Caltech Submillimeter Telescope (CSO)

UH Hilo Educational Telescope

Pu`u Poliahu

Summit 13,796 feet

Pu`u Wekiu

13300
13400
13500
13600
13500
13400
13250
13550
13500
13450
13200
13300
13400
13500
13600
13700

Mauna Kea Summit

Hilo

Hawaii "The Big Island"

N

0   400   800   1200 feet

Contour intervals of 50 feet

UNIVERSITY OF HAWAI'I INSTITUTE FOR ASTRONOMY

MAUNAKEA TELESCOPES

NAME	SIZE	DATE	WEB SITE
University of Hawaiʻi 2.2-meter Telescope	2.2m	1970	www.ifa.hawaii.edu
Canada-France-Hawaiʻi Telescope (CFHT)	3.6m	1979	www.cfht.hawaii.edu
NASA Infrared Telescope Facility (IRTF)	3.0m	1979	irtfweb.ifa.hawaii.edu
United Kingdom Infrared Telescope (UKIRT)	3.8m	1979	www.jach.hawaii.edu/UKIRT
Caltech Submillimeter Observatory (CSO)	10.4m	1987	www.cso.caltech.edu
James Clerk Maxwell Telescope (JCMT)	15.0m	1987	www.jach.hawaii.edu/JCMT
Very Long Baseline Array (VLBA)	25.0m	1992	www.vlba.nrao.edu
Keck I Telescope	10.0m	1993	www.keckobservatory.org
Keck II Telescope	10.0m	1996	www.keckobservatory.org
Gemini North Telescope	8.0m	1999	www.gemini.edu
Subaru Telescope	8.2m	1999	www.subarutelescope.org
Submillimeter Array (SMA)	8x6m	2002	www.cfa.harvard.edu/sma
University of Hawaiʻi 0.9m Educational Telescope	0.9m	2010	www.astro.uhh.hawaii.edu
Thirty Meter Telescope (TMT)	30.0m	2022	www.tmt.org

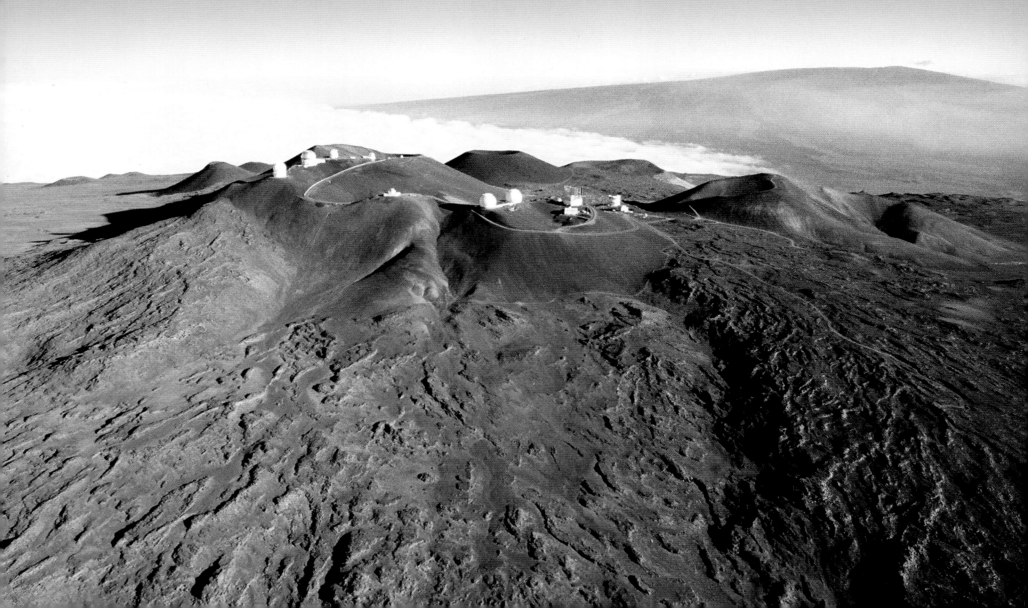

Mauna Kea is truly a very special place.
At the summit we can drink in the
wisdom and beauty of the gods: the
heavens full of stars, the Milky Way
and myriad worlds of mystery.

—WALTER STEIGER

A PLACE LIKE NO OTHER

HOW DID A DORMANT VOLCANO in the middle of the Pacific Ocean become home to the world's most powerful collection of telescopes? It's all about location.

We live at the bottom of an ocean of air. Clouds, wind, humidity, and artificial lighting degrade our view of the heavens, distorting—and in some cases blocking—light from planets, stars, and galaxies. Because Maunakea's 13,796-foot (4,205-meter) summit towers over nearly half our planet's atmosphere and the remaining thin, dry air overhead is exceptionally calm, telescopes there can obtain images of unsurpassed clarity. The result is Earth's best window on the universe.

There are, of course, other superb astronomical sites around the globe. Chile's Atacama Desert, sandwiched between the Pacific Ocean and the Andes Mountains, is home to several major observatories. The Canary Islands off Africa's western coast, Mexico's Baja California peninsula, and even sites in Antarctica are also places with exceptionally good conditions for astronomical observations. But geography has blessed Maunakea with a unique combination of traits found nowhere else, a consideration that led the Thirty Meter Telescope to choose this mountain over all others as its future home.

A Hawaiian proverb says, *O Hawai'i no ka 'āina maika'i*, which means "there's no finer land than Hawai'i." As the following pages show, Maunakea is truly one of the most glorious places on this or any island, providing stunning views of the world below and of the universe above.

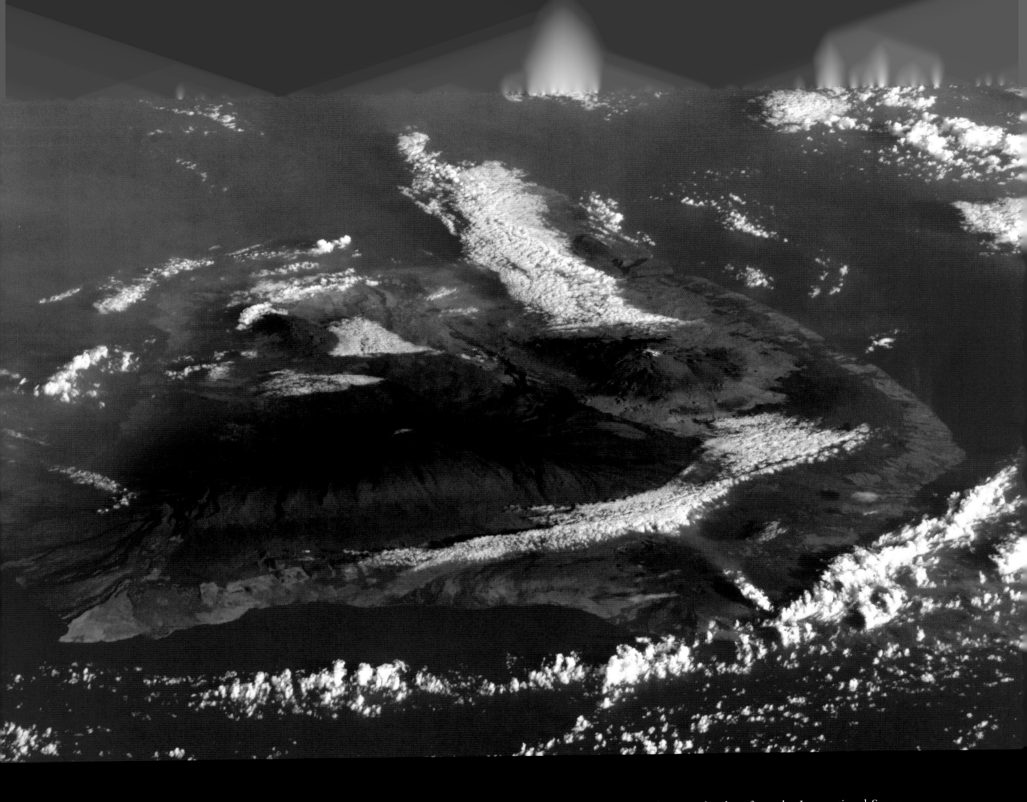

Astronomy Island

The island of Hawai'i is seen in this 2002 photograph taken from the International Space Station while it orbited Earth at an altitude of 245 miles (394 kilometers). The island is dominated by Maunaloa to the left and Maunakea to the right. Maunakea is the slightly taller of the two mountains and has much steeper slopes. A plume of smoke from Kīlauea volcano can be seen near the bottom right as lava flows into the sea. PHOTO: NASA

American author Mark Twain visited the Hawaiian Islands in 1866, describing them as "the loveliest fleet of islands that lies anchored in any ocean." Their diversity left a lasting impression on him. He wrote, "If you want snow and ice forever and ever, and zero and below, build on the summit of Mauna Kea."

This photo shows the snow-capped summit of Maunakea as seen in 2003 from the International Space Station while it orbited 247 miles (397 kilometers) above Hawai'i.
PHOTO: NASA

Fire and Ice

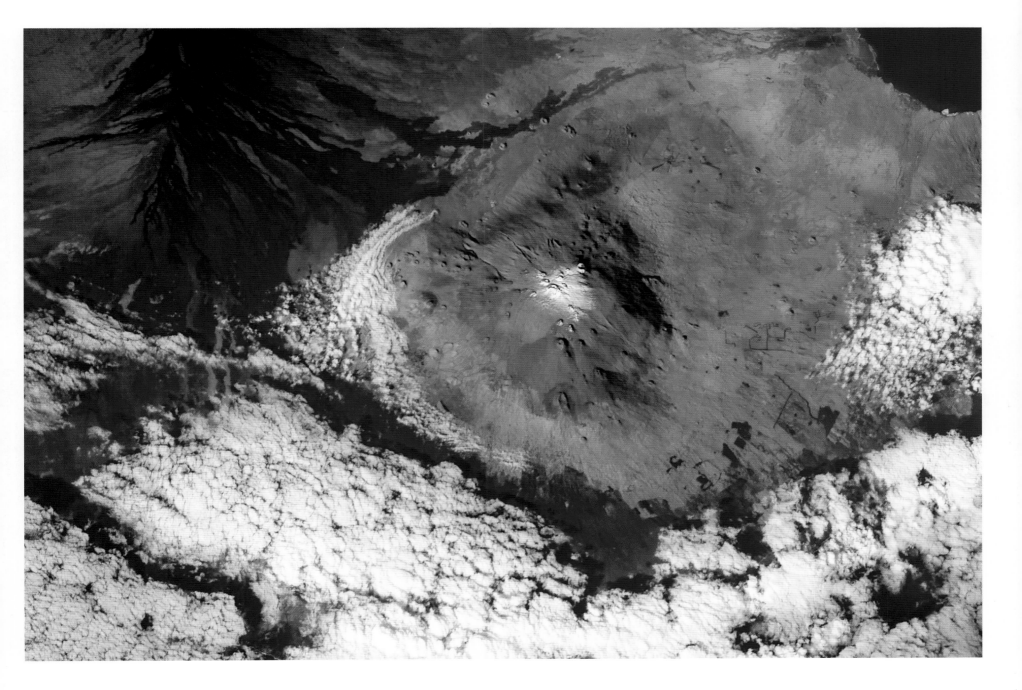

Land of Volcanoes

Like all the volcanoes that make up the Hawaiian Islands, Maunakea was born of fire. Sometime in the distant past, roughly a million years ago, the mountain was created by a volcanic hot spot on the ocean floor that belched molten lava from Earth's interior. As the newborn Maunakea grew, its crest slowly emerged from the ocean's watery womb and continued rising until it pierced the clouds. Carried along by our planet's constantly shifting crust, the massive mountain—which today is bigger than the islands of Oʻahu and Maui—gradually drifted away from its parental hot spot and grew no more.

The Big Island of Hawaiʻi is home to several volcanoes. Here Hualālai's 8,271-foot (2521-meter) summit pokes through the clouds in the distance during a golden sunset, with Maunakea's lovely Puʻu Poliʻahu cinder cone in the foreground. PHOTO: MICHAEL WEST

Stargazing in Hawai'i goes back nearly two thousand years. "The ancient Hawaiians were astronomers," wrote Queen Lili'uokalani, Hawai'i's last reigning monarch, in 1897.

The Hawaiian people looked to the skies for both spiritual and secular guidance in their lives. Crops were planted, fish were caught, battles were fought, and religious festivals were celebrated according to the phases of the moon and the seasonal positions of the stars in the sky. Hawaiian astronomers, called *kilo hōkū* or "star watchers," were among the most revered members of Hawaiian society.

The *Kilo Hōkū*

Here a recently constructed traditional Hawaiian *lele* sits atop an old lava flow, with snow-capped Maunakea in the background. The mountain is considered a sacred site by some Native Hawaiians, who come to place offerings to Poli'ahu, the goddess of snow, and her sister Lilinoe, the goddess of mist. Some say the snow that often covers the mountain's peak during winter months is the source of its name; "mauna kea" means "white mountain." Others suggest that Maunakea is a shortened form of Ka Mauna a Wākea, or "Wākea's Mountain." According to ancient lore, the island of Hawai'i was the firstborn child of Wākea and Papa-hānau-moku, the Sky Father and Earth Mother. PHOTO: BARBARA SCHAEFER

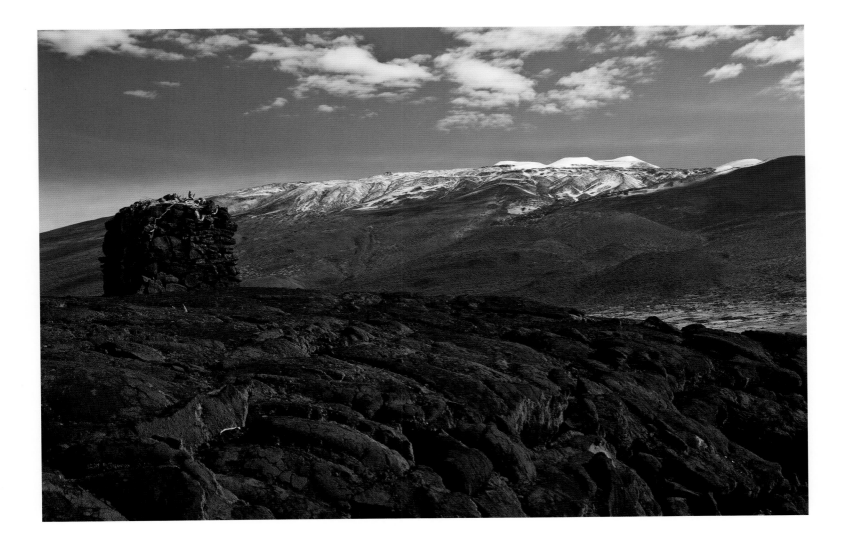

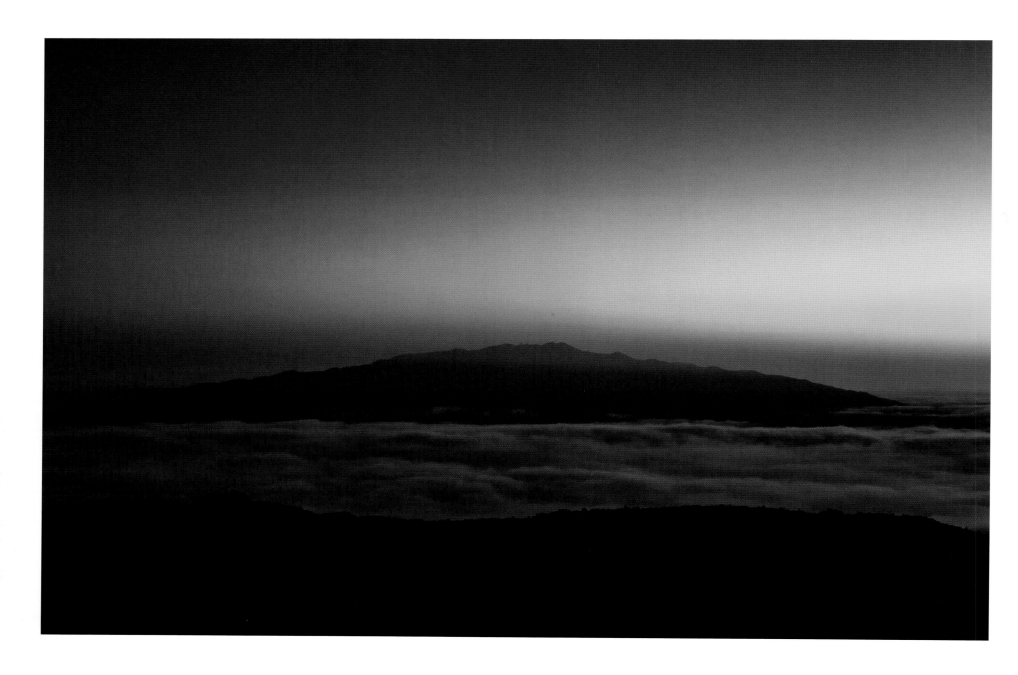

New Day Rising

The first light of dawn illuminates Maunakea as seen from the summit of its sister mountain, Maunaloa. PHOTO: MICHAEL WEST

Maunakea rises in the distance as seen from Hilo Bay at sunset. Northeast trade winds bring moist air that condenses as it rises up the mountain's slopes, making Hilo one of the rainiest cities in the world. PHOTO: MICHAEL WEST

The View
from Hilo Bay

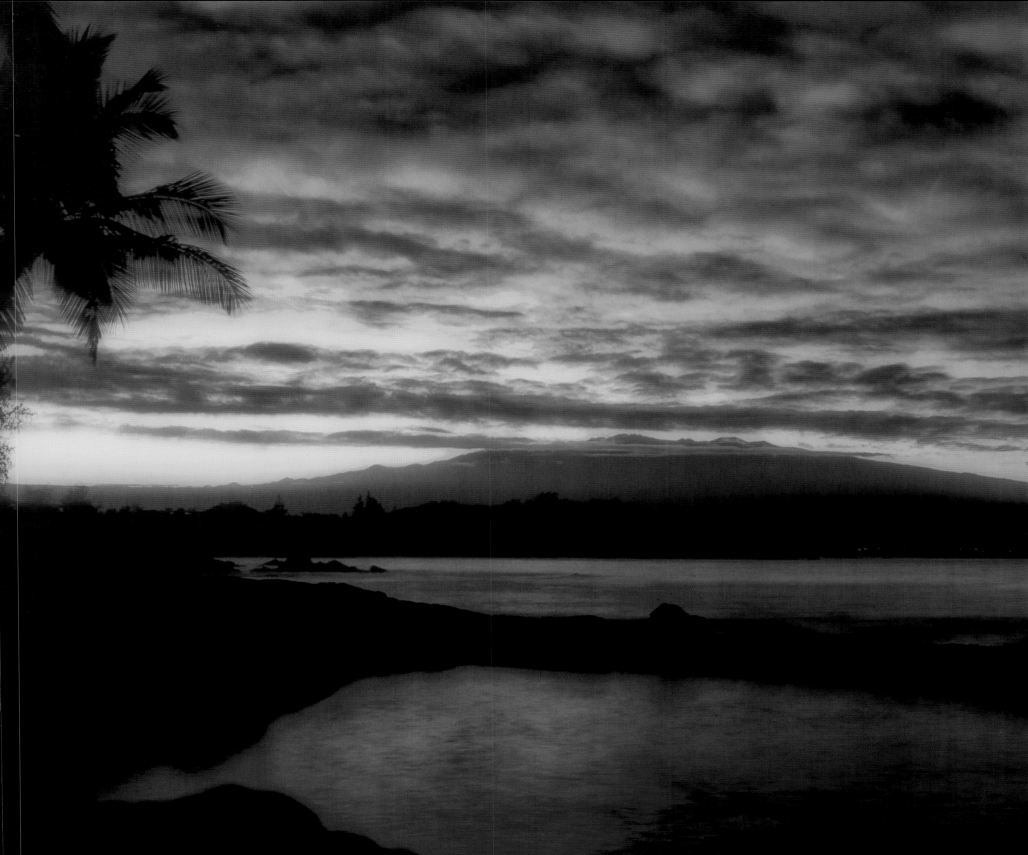

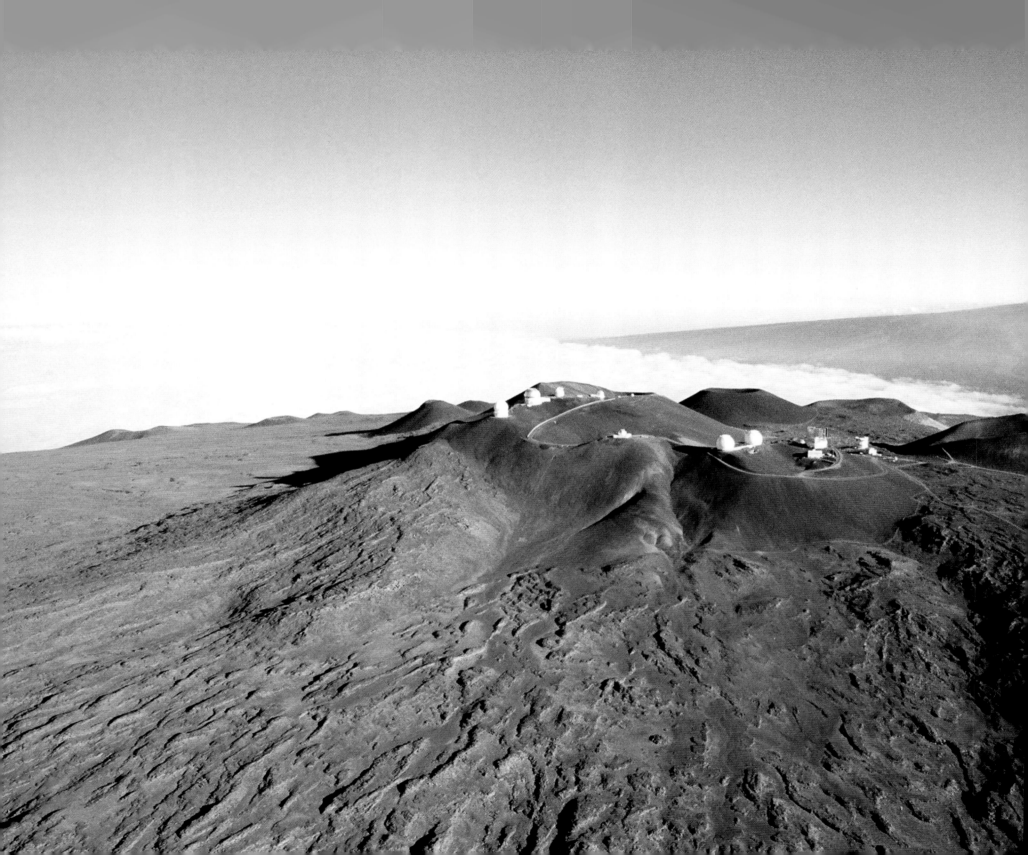

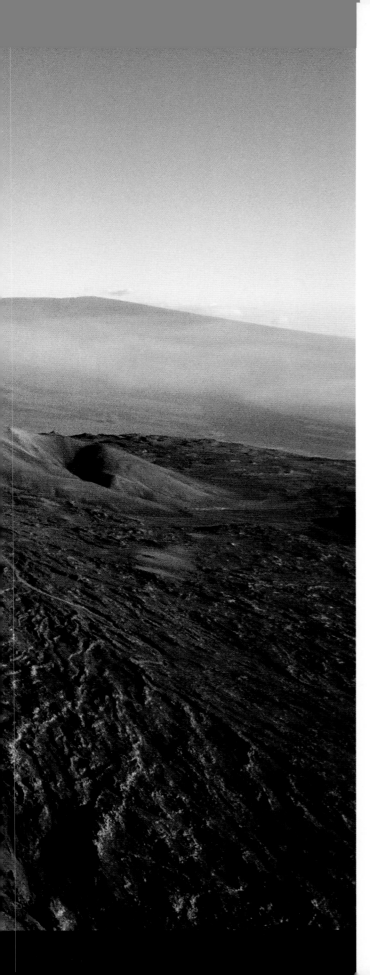

The Scars of Time

The island of Hawai'i is home to five volcanoes—Maunakea, Maunaloa, Kohala, Hualālai and Kīlauea—with a sixth named Lō'ihi erupting below the sea and slowly creating the next Hawaiian island. Kohala and Maunakea are the only two of these volcanoes that have not erupted during recorded history. But Maunakea's history has been anything but placid. Carved by glaciers and shaped by lava flows to which no humans were witness, the mountain still bears the scars of its youthful past. PHOTO: SUBARU TELESCOPE/NAOJ

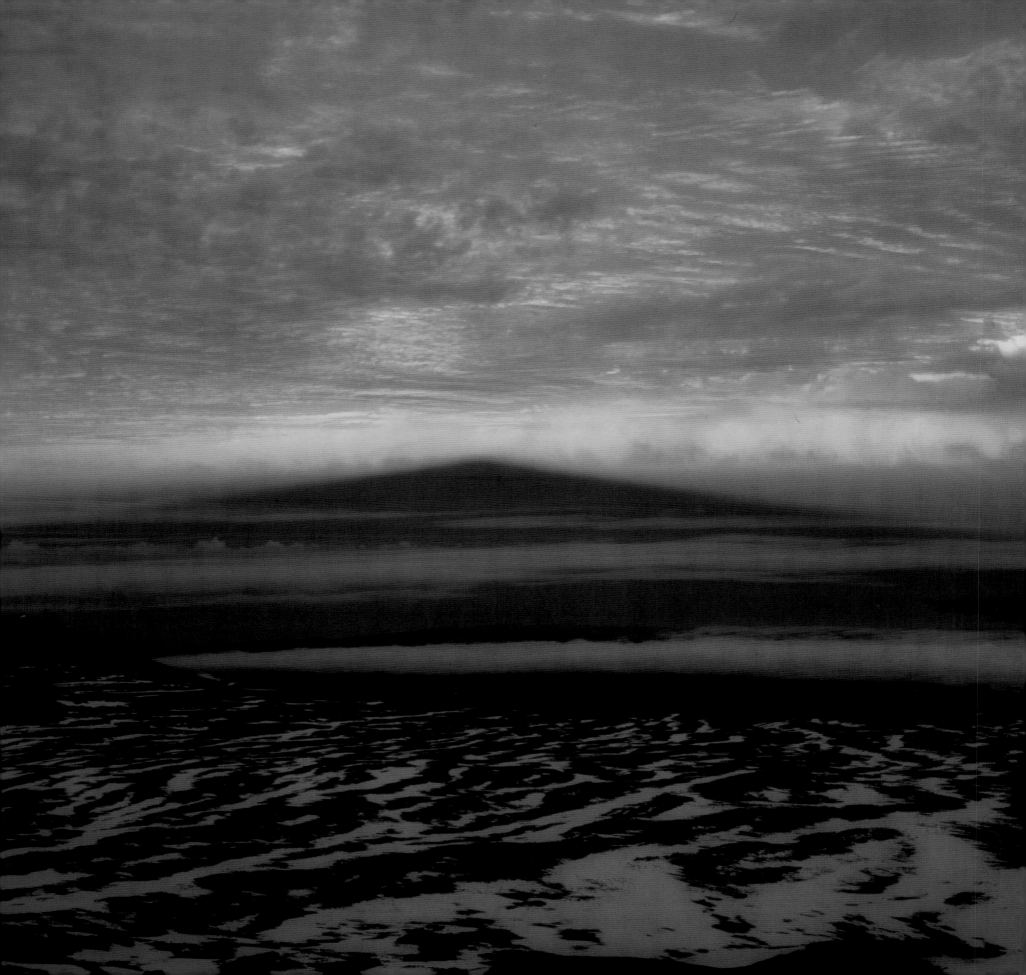

The Shadow
of Maunakea

As the sun sets, Maunakea's shadow stretches across Earth's upper atmosphere and the clouds that lie below the mountain's summit. PHOTO: MICHAEL WEST

Despite its tropical location, Maunakea has experienced ice ages at least four times in its history, the last one about 12,000 years ago. Geological evidence suggests that glaciers covering the summit reached thicknesses of 330 feet (100 meters) or more. As the glaciers slowly retreated, they scraped and sculpted the land beneath them, leaving behind piles of stones as well as scars etched into the mountain's surface. Two of Maunakea's most prominent cinder cones, Puʻu Wēkiu and Puʻu Poliʻahu, seen here in the background, have unusually steep slopes and distinctive coloring that resulted from glaciers scraping away their sides and exposing their interiors. PHOTO: MICHAEL WEST

Maunakea's Ice Ages

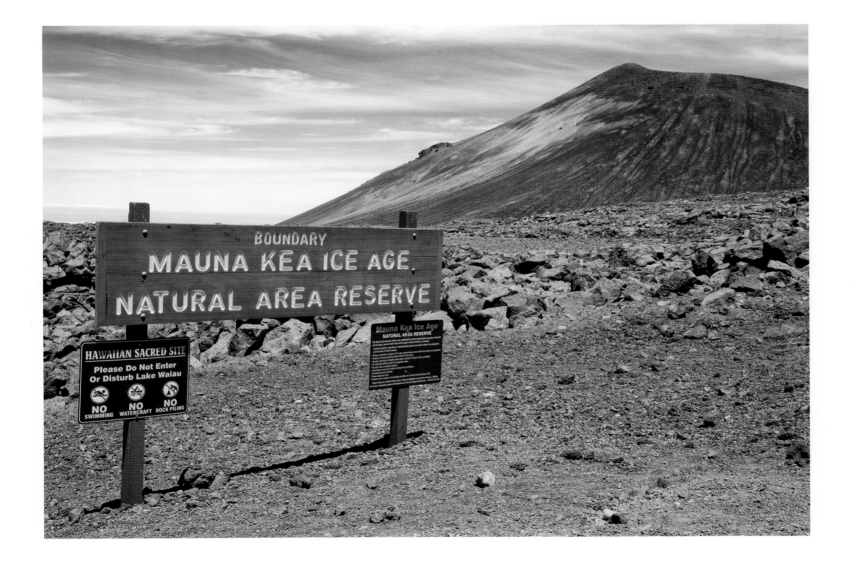

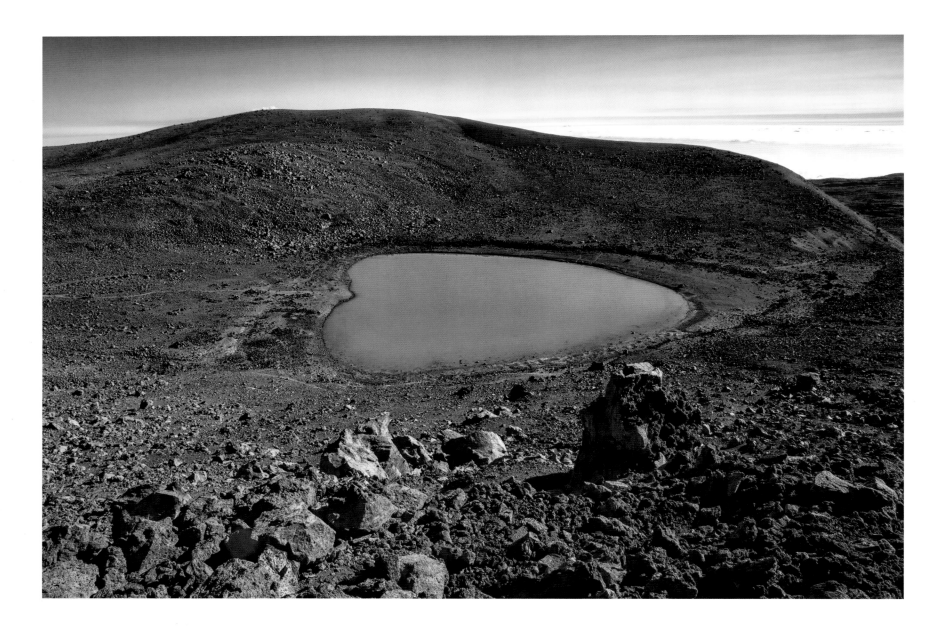

Lake Waiau

At an altitude of 13,020 feet (3,968 meters), Maunakea's enchanting Lake Waiau is Hawai'i's only alpine lake. Its name means "swirling waters" in Hawaiian, although the lake is usually quite placid. Formed by a glacier that swept across Maunakea some 30,000 years ago, the lake is replenished today by occasional rain and melted snow. Silt, clay, and ash have accumulated at the bottom of the lake to create an impermeable layer that prevents the water from seeping into the porous rock below. Lake Waiau's vivid green color comes from algae that thrive in its pristine waters. Ancient Hawaiians believed that the lake was bottomless and considered it a sacred place, as do some Hawaiians today.

Hiking to the lake in 1922, Lawrence Daingerfield wrote in his journal, "Taking our last look across the lake, we saw the image of fair Venus shimmering light across the tiny, rippling waves. A thousand jewels glittered in the reflected light." PHOTO: MICHAEL WEST

Much of Maunakea's success as an astronomical site can be attributed to geography. Surrounded by thousands of miles of open ocean, with no large landmasses nearby to roil the air, winds glide smoothly across the mountain's summit, producing an especially stable atmosphere over the observatories that yields some of the sharpest astronomical images obtainable anywhere on Earth. These exquisite atmospheric conditions, together with the high percentage of cloud-free nights and minimal light pollution on the sparsely populated island of Hawai'i, have made Maunakea the premiere astronomical site in the world.

The Big Island is seen in this photograph taken onboard the International Space Station on August 13, 2008, from an altitude of 217 miles (349 kilometers). PHOTO: NASA

A S p l e n d i d   I s o l a t i o n

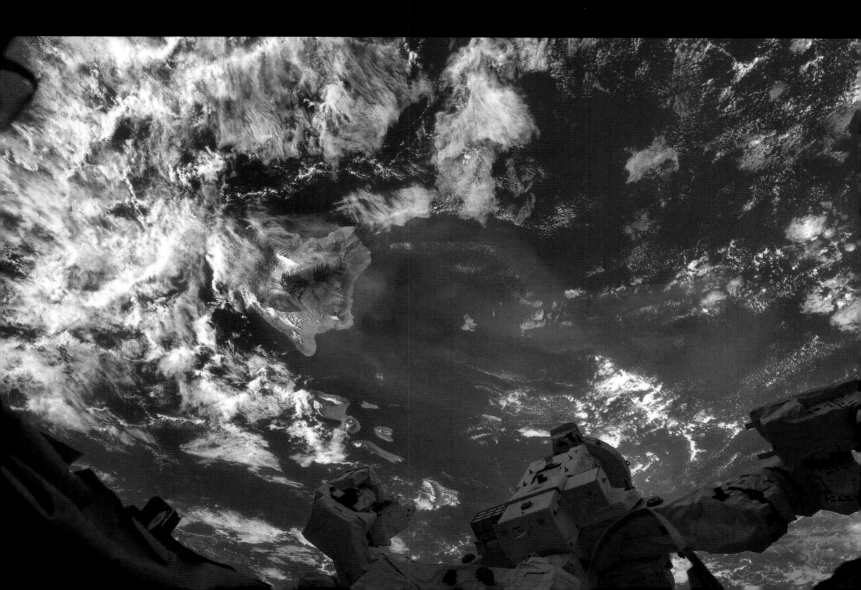

Pu'u Wēkiu— The Summit of Maunakea

The highest point on Maunakea—and in all of Hawai'i—is Pu'u Wēkiu, a cinder cone that reaches an elevation of 13,796 feet (4,205 meters). PHOTO: BARBARA SCHAEFER

A Landscape Dotted with Cinder Cones

Geologists have identified nearly one hundred cinder cones on Maunakea's slopes, a reminder of the mountain's active past. These cinder cones formed when lava fragments were forcefully ejected into the air during volcanic eruptions and rained back down to form mounds of ash and cinders. Many cinder cones, like this one, have a bowl-shaped crater at the top. PHOTO: MICHAEL WEST

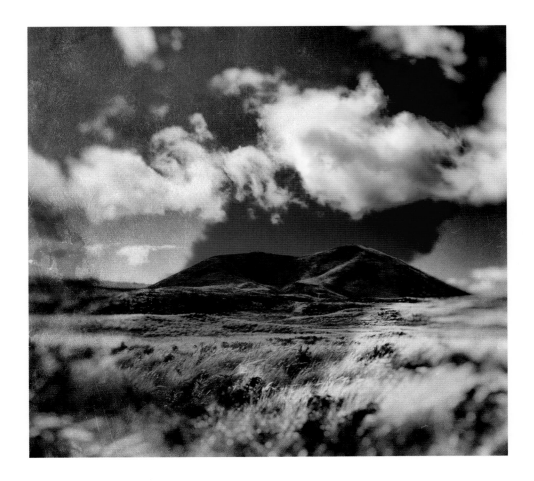

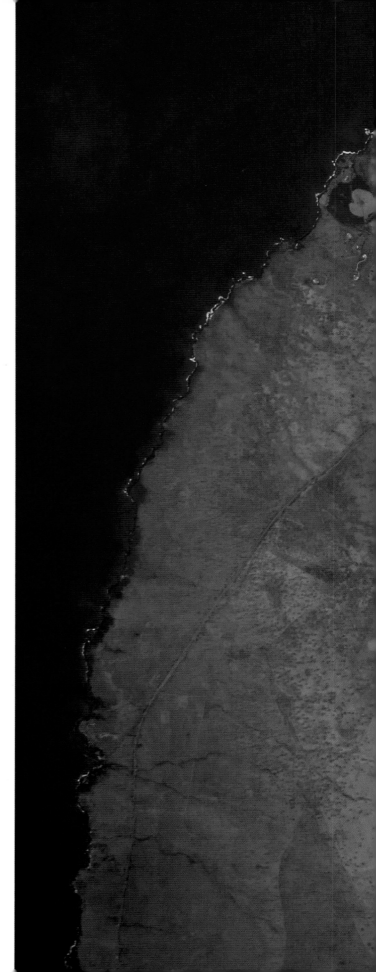

Wet and Dry

Maunakea's preeminence as an astronomical site owes much to its unique location and geography. The mountain plays a major role in shaping the island's weather, as moist trade winds from the northeast rise up Maunakea's slopes and condense into clouds and rain. Because of this, the eastern side of the mountain is a lush rain forest, while the western side of the island has desertlike conditions. Hilo is one of the rainiest cities in the United States, receiving anywhere from 120 to 180 inches of rain a year—three to four times that of Seattle.

These contrasting conditions are evident in this image of the northern part of the Big Island as seen from the International Space Station. PHOTO: NASA

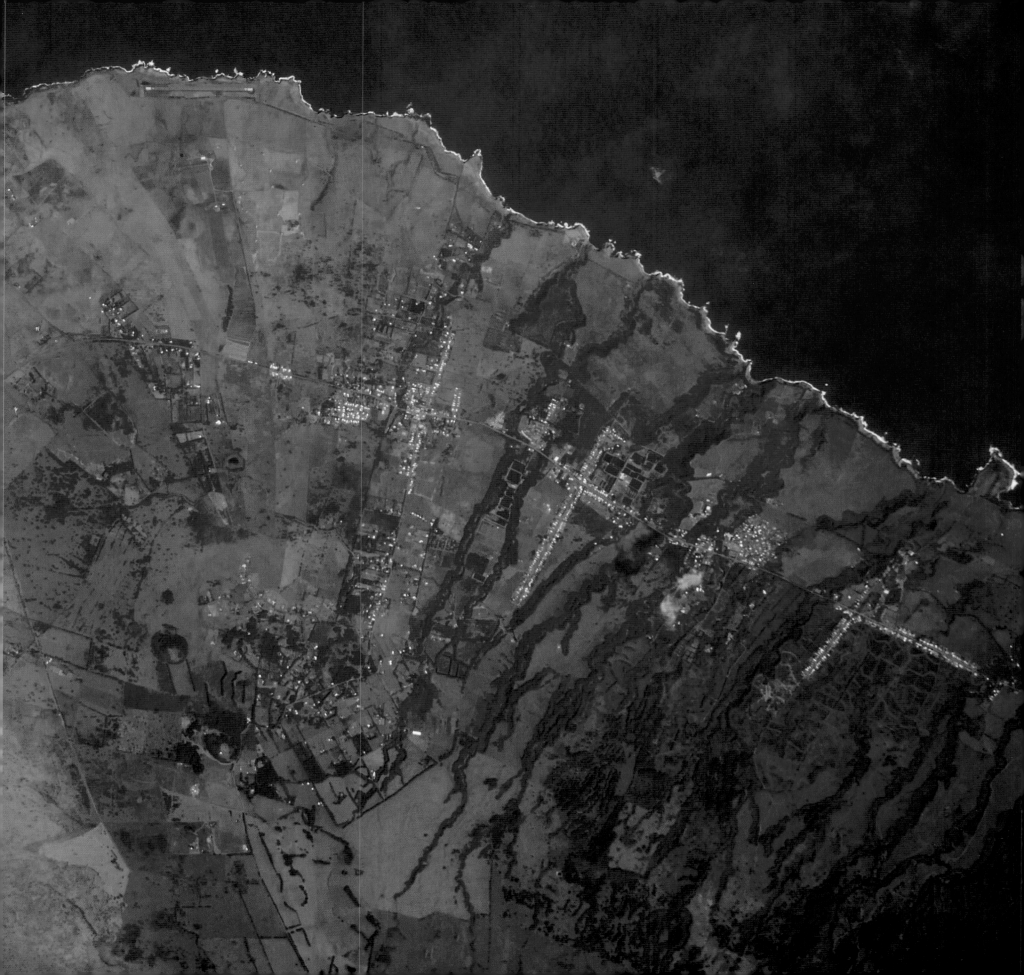

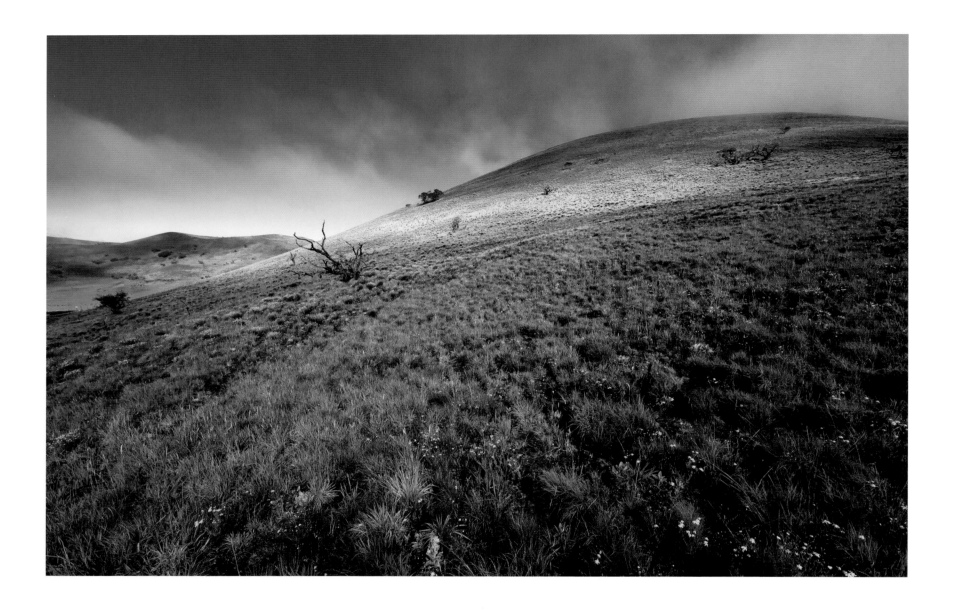

Maunakea's Fertile Flanks

The Island of Hawai'i is home to one of the world's most varied environments. Here you'll find deserts, rain forests, alpine conditions, and other climatic regions.

These extremes are especially pronounced on Maunakea. Mark Twain, who visited Hawai'i as a young reporter in 1866, remarked that from the top of Maunakea, "you can look down upon all the climates of the earth." The contrast between the mountain's barren heights and its fertile lower elevations is especially stunning. Maunakea's slopes are home to a diverse ecosystem that includes native *māmane* trees, rare silversword plants, endemic *palila* birds, grazing cattle, wild pigs, and sheep. PHOTO: MICHAEL WEST

A Garland
of Stars over
Puʻu Wēkiu

The Milky Way hangs like a garland of
stars over Puʻu Wēkiu, the highest point on
Maunakea. No one knows for certain whether
the ancient Hawaiians made astronomical
observations from Maunakea. Perhaps the
kilo hōkū, Hawaiian star watchers, came to
the tallest mountain in the Pacific to discern
signs in the starry skies from its vantage point
closest to the heavens. Or perhaps the summit
was considered too divine for such human
activities. PHOTO: MICHAEL WEST

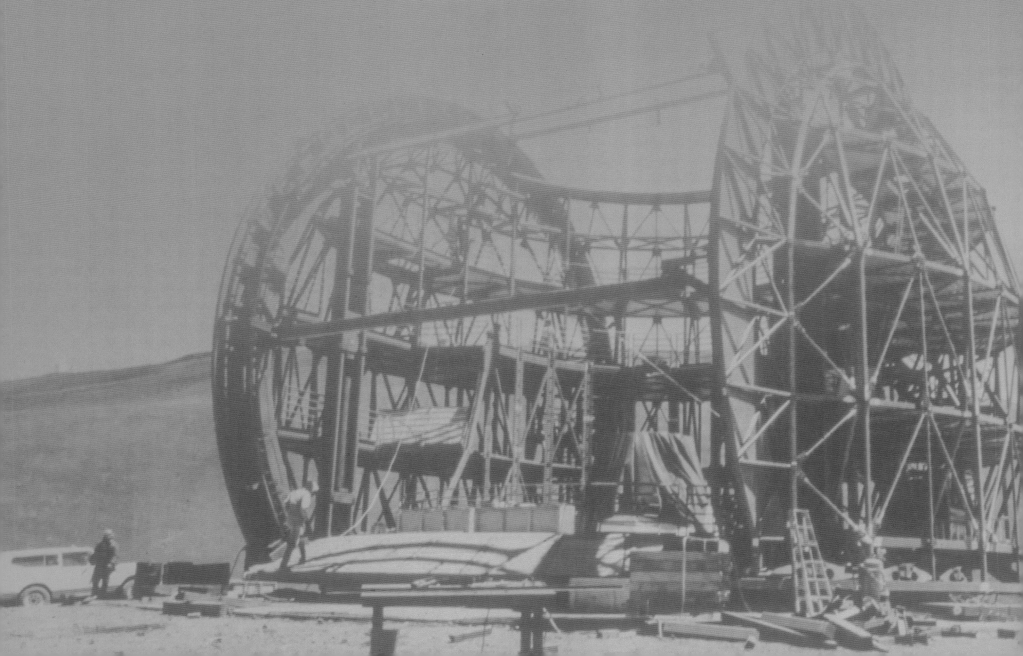

If you want to build a ship, don't drum up people together to collect wood and don't assign them tasks and work, but rather teach them to long for the endless immensity of the sea.

—ANTOINE DE SAINT-EXUPÉRY

FROM DREAM TO REALITY

LIKE FISHERMEN CASTING THEIR NETS into the sea, astronomers snare the light of planets, stars, and galaxies with telescopes. Ever since Galileo first turned his primitive spyglass toward the heavens in 1609, making visible what was once invisible, no other invention has revolutionized our view of the universe—or our place in it—more than the telescope.

The small sizes and simple designs of early telescopes meant that a single person with enough ingenuity and resources could make one. But just as a larger net catches more fish, a larger telescope catches more light. As telescopes grew in size and complexity, so did the effort required to build and operate them. By the nineteenth century, telescopes had become behemoths housed in observatories, attended to by multiple caretakers.

Modern astronomy on Maunakea began with Gerard Kuiper's dream of finding the best site in the world to build a new observatory, a quest that brought him to the island of Hawai'i in 1964. There he found what he'd been searching for: a place of pristine skies where astronomical images of unrivaled clarity could be obtained, a place he called "a jewel."

But it takes more than dreams to build an observatory. Years of planning, fundraising, cajoling, constructing, and testing must all come together before a newborn telescope can open its eye to the night sky. A multitude of challenges—some anticipated, some not—have to be overcome along the way.

The thirteen telescopes perched atop Maunakea represent the collective dreams and labors of many people, some of them forgotten, who saw this mountain as something special—a gateway to the stars—and devoted years of their lives to turn that vision into reality. It's a dream that continues today with plans for the next-generation Thirty Meter Telescope, whose light-gathering power will exceed that of more than one hundred Hubble Space Telescopes, bridging heaven and earth like never before.

A Hawaiian proverb says, *E lauhoe mai na wa'a pae aku i ka 'āina*, which means "Everybody paddle the canoes together and the shore is reached." On quiet nights on Maunakea, as the telescopes stare into the starry skies, one can almost hear the faint sound of hundreds of oars being dipped into the waters of the cosmic ocean, answering an ancient call beckoning us back to the stars.

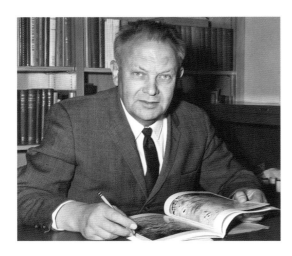

Kuiper's Dream

Modern astronomy on Maunakea began in 1964, when University of Arizona astronomer Gerard Kuiper came to Hawai'i in search of the best astronomical site in the world. There, with funding from NASA, he hoped to build an observatory to capture the most magnificent views of the heavens obtainable anywhere. After testing nighttime conditions on Maunakea, Kuiper declared enthusiastically, "This mountaintop is probably the best site in the world—I repeat in the world—from which to study the Moon, the planets and the stars. . . . It is a jewel!"

To his great disappointment, Kuiper never realized his dream of building a telescope on Maunakea. NASA chose to fund a competing proposal from the University of Hawai'i to build an 88-inch telescope on the mountain, which was completed in 1970.

Never one to give up, Kuiper always held out hope that an even better astronomical site than Maunakea might exist somewhere, and he continued his quest to find it. But he never did. Kuiper died in Mexico in 1973 while scouting possible sites for new observatories there. The telescopes on the summit of Maunakea remain his legacy, testament to one man's dream of finding the best place on Earth to observe the heavens. PHOTO: AIP EMILIO SEGRÈ VISUAL ARCHIVES, PHYSICS TODAY COLLECTION

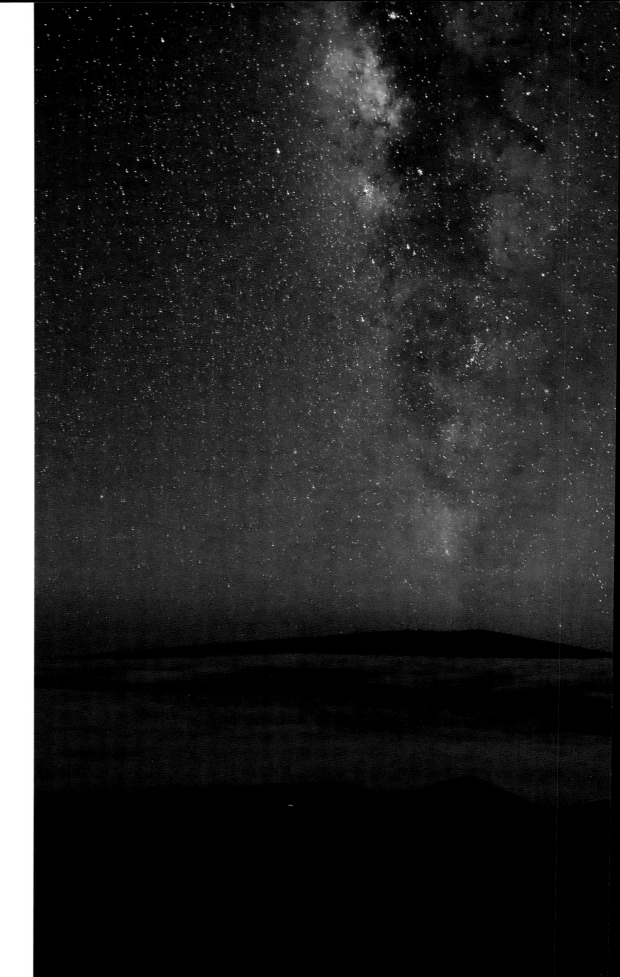

Why Not Maunaloa?

When Gerard Kuiper was searching for sites to build a new telescope in the 1960s, he first explored Haleakalā, the highest peak on the island of Maui. The University of Hawai'i had built a solar telescope there in 1961 and found the atmosphere to be remarkably stable during the day. Nighttime tests by Kuiper and his assistants confirmed that astronomical images of extraordinary clarity could be obtained there. Unfortunately, clouds and fog often sweep over Haleakalā at night, blocking views of the heavens, and so Kuiper turned his attention elsewhere.

The summits of Maunakea and Maunaloa—both almost 4,000 feet higher than Haleakalā—could often be seen poking through the clouds over on the neighboring island of Hawai'i. Maunaloa, nearly as high and more accessible than Maunakea, might seem a natural choice except for one major drawback: it's one of the most active volcanoes in the world today and hence not a safe place to build an observatory. Because Maunakea hasn't erupted for thousands of years, Kuiper chose it for further site testing. Today, it's home to the world's most powerful collection of telescopes. PHOTO: MICHAEL WEST

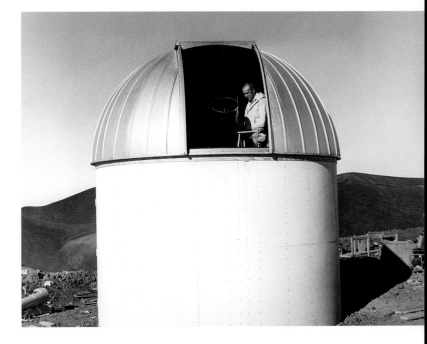

"A Sky Wonderful with Stars"

Recounting her ascent of Maunakea in 1873, Victorian-era travel writer Isabella Bird wrote, "The mist as usual disappeared at night, leaving a sky wonderful with stars."

Astronomer Alika Herring, who grew up in Hawai'i, sits in the first observatory on Maunakea in 1964. Herring, a renowned telescope maker who built the small telescope seen here, did most of the site testing on Maunakea as part of Gerard Kuiper's team. He considered it the best astronomical site he'd ever seen. PHOTO: COURTESY OF THE LUNAR AND PLANETARY LABORATORY, UNIVERSITY OF ARIZONA

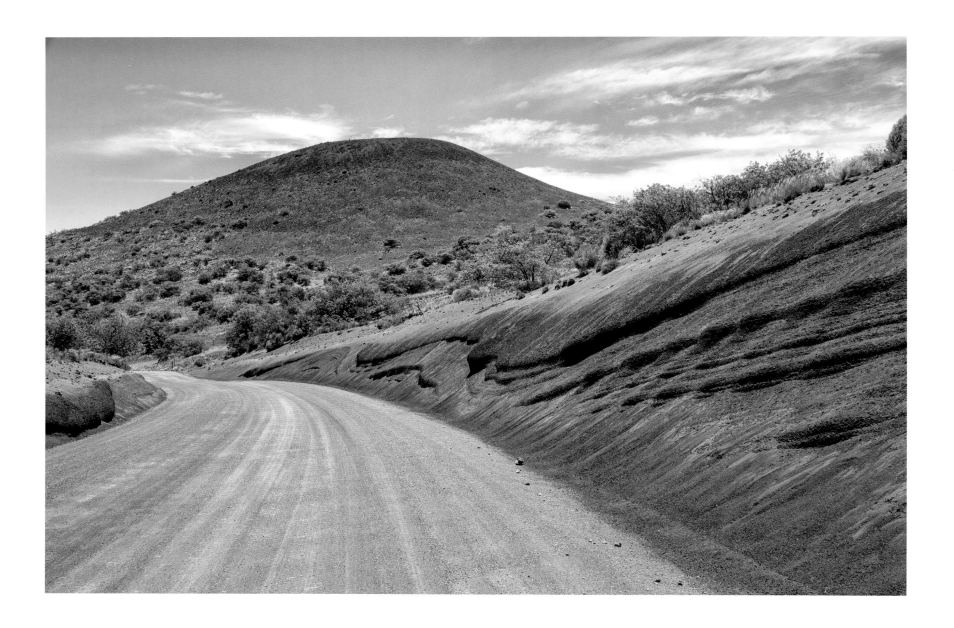

The Road to Maunakea

Construction of a road to the summit of Maunakea began in April 1964 and was completed within a month. From Hale Pōhaku, located about halfway up the mountain, the journey to the top of Maunakea is 8 miles (13 kilometers) long, much of it steep and winding. The first few miles are unpaved and very rough in places. The road is paved near the summit to minimize dust blowing into the telescopes. Road crews must resurface the unpaved portion of the road weekly. PHOTO: MICHAEL WEST

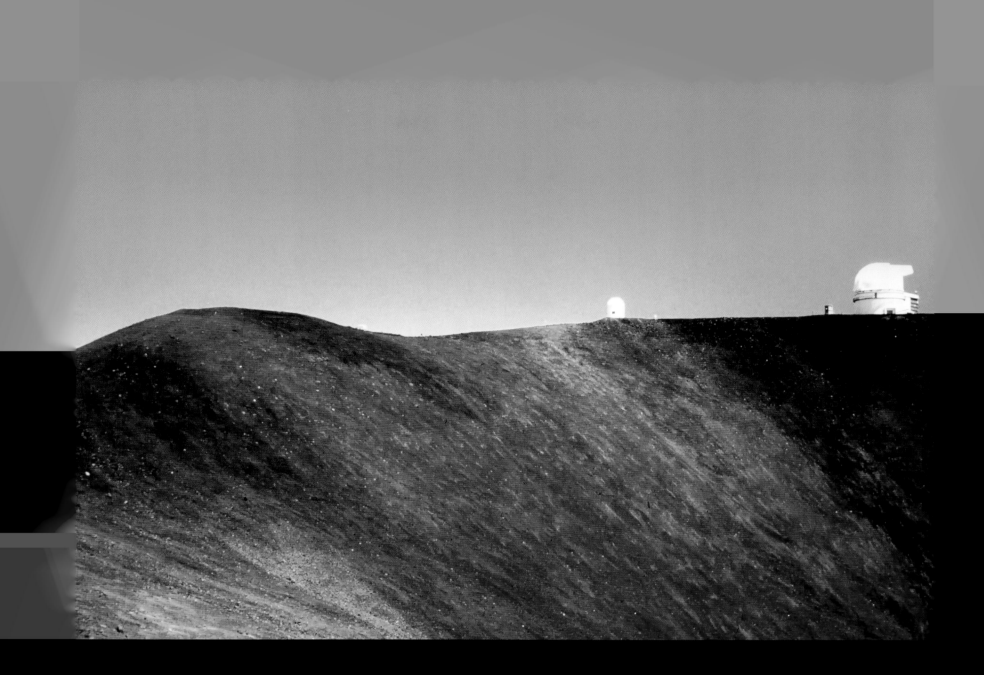

The first telescope to begin operation on Maunakea was a 24-inch U.S. Air Force telescope that was used to identify man-made objects in space. This was soon followed by a second 24-inch telescope built by Lowell Observatory as part of the International Planetary Patrol, a network of six telescopes around the globe to study objects in our solar system.

This photograph shows the Planetary Patrol telescope (center) and the University of Hawaiʻi 88-inch telescope (right) that saw first light in 1970. The site of the Canada-France-Hawaiʻi Telescope (CFHT), still several years in the future, is located on the far left. The Planetary Patrol telescope was removed in 1995 to make room for the Gemini North telescope. PHOTO: INSTITUTE FOR ASTRONOMY ARCHIVES, UNIVERSITY OF HAWAIʻI

Future Site of the Canada-France-Hawaii and Gemini North Telescopes, Circa 1970

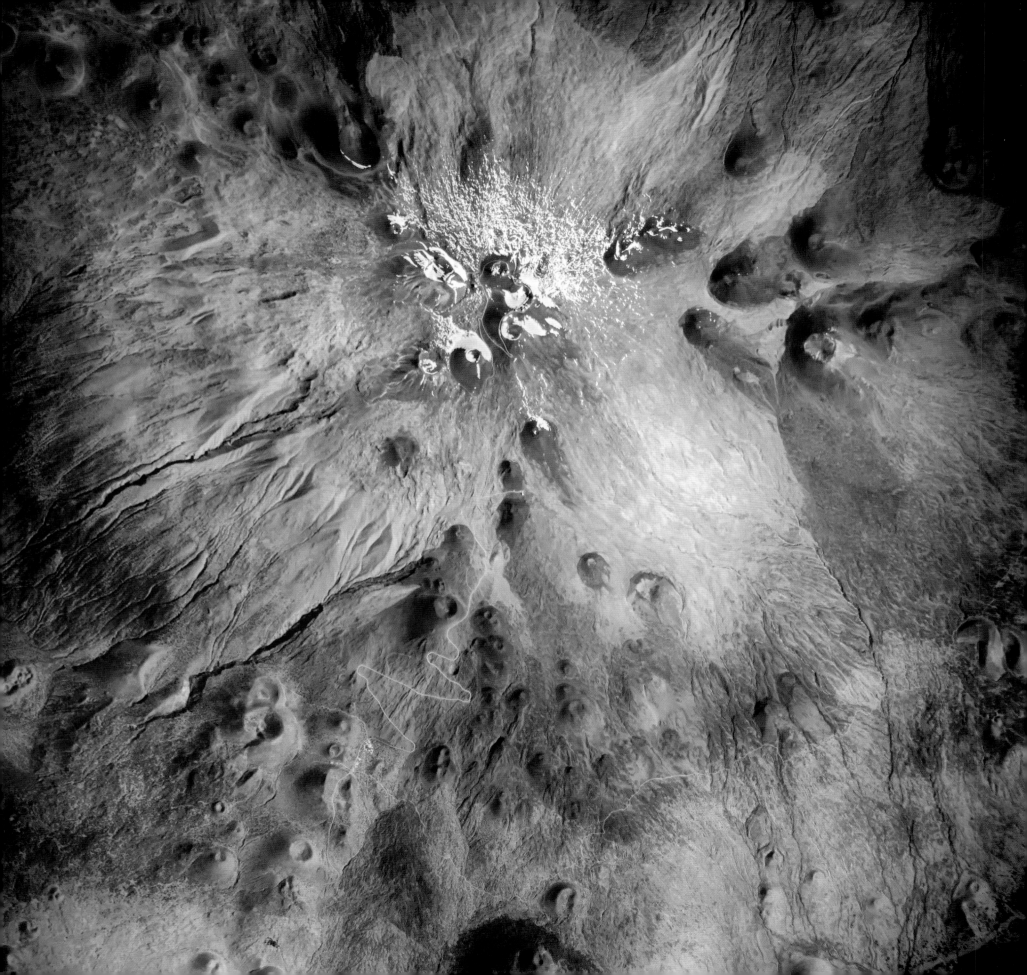

The View from Space

Although it might look like the surface of Mars, this photo is actually the summit of Maunakea as photographed by astronauts on the International Space Station in 2009. The winding road to the summit can be seen stretching from the bottom to the summit near the middle of the photograph. PHOTO: NASA

Steiger's Vision

If Gerard Kuiper was the father of modern astronomy on Maunakea, then Walter Steiger was its grandfather.

Steiger first came to the Islands as a soldier during World War II, an experience that changed his life forever. After completing his PhD in physics at the University of Cincinnati, he happily accepted an offer of a professorship at the University of Hawai'i.

Recognizing the great potential of Hawai'i's natural resources for scientific research, he established the first modern astronomical observatory in the state, beginning with a temporary facility to monitor solar activity from Makapu'u Point on the island of O'ahu in 1957. This was followed by a permanent solar observatory in 1961, built on the summit of Healakalā on the island of Maui.

Steiger's success encouraged others, most notably Gerard Kuiper and John Jefferies, to pursue the possibility of building astronomical observatories in Hawai'i, paving the way for Maunakea to become one of the leading centers of astronomy in the world.

Steiger is seen here in front of an image of the sun that was taken in 1958 with the telescope at Makapu'u Point. Many who knew him were touched by his aloha spirit.

PHOTO: MICHAEL WEST

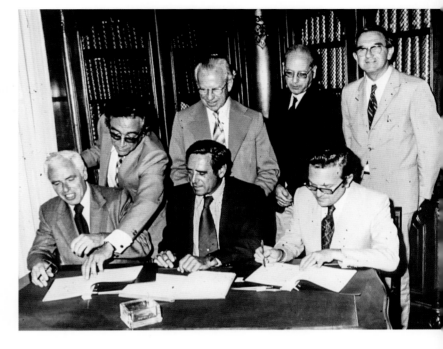

Jefferies' Determination

The University of Hawai'i has risen to become one of the leading centers of astronomical research in the world today thanks in large part to the efforts of John Jefferies.

Jefferies, originally from Australia, was the founding director of the University of Hawai'i's Institute for Astronomy. He arrived in Hawai'i in 1964, drawn by the opportunity to do research using the University of Hawai'i's new solar observatory on Haleakalā, the highest peak on the island of Maui.

At that time, NASA was interested in funding a telescope in Hawai'i and was considering proposals from well-established astronomy departments at Harvard and the University of Arizona. Seizing the opportunity, Jefferies suggested that NASA instead choose the fledgling astronomy program at the University of Hawai'i to be its partner. Within a month of receiving Jefferies' proposal, NASA agreed, and in July 1965 the university was awarded $3 million to construct a new observatory on Maunakea.

In this image, Jefferies (seated, center) joins colleagues from Canada and France to sign the tripartite agreement that led to construction of the Canada-France-Hawai'i Telescope. PHOTO: CANADA-FRANCE-HAWAI'I TELESCOPE

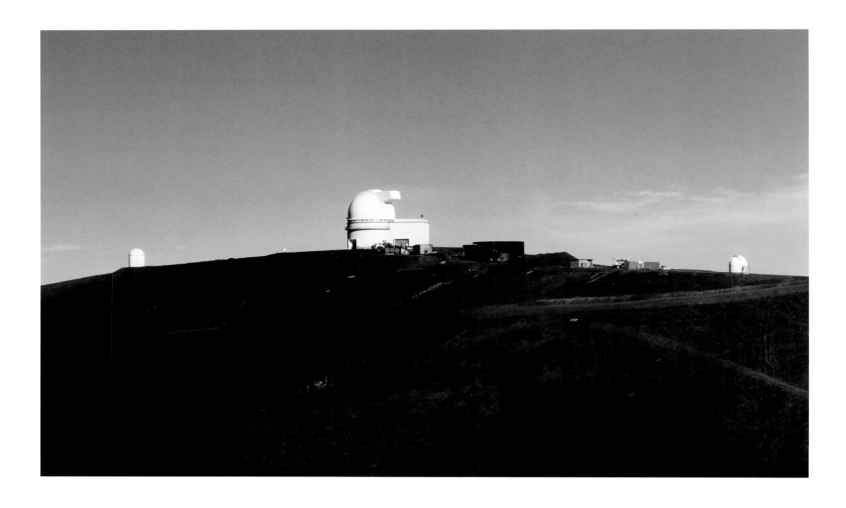

The Institute for Astronomy

The University of Hawai'i's Institute for Astronomy (IfA) was founded in 1967 to manage the Haleakalā Observatories on Maui and guide the development of the Maunakea Observatories on Hawai'i Island. Today, it has a total staff of more than three hundred, including approximately eighty PhDs from twenty-six different countries, making it one of the largest university astronomy programs in the world. Most IfA staff are based at the university's Mānoa campus on the island of O'ahu, with others located at satellite offices on the Big Island and Maui.

IfA astronomers conduct forefront research and have access to a guaranteed share of time on all telescopes on Maunakea and Haleakalā. Pioneering research done by IfA faculty has been recognized with numerous prestigious awards. The IfA has also been a leader in developing new technologies for ground-based and space-based observatories, most notably making innovative contributions to the development of adaptive optics and infrared array detectors. As part of the University of Hawai'i system, the IfA also contributes in important ways to the university's educational mission and has produced more than 150 PhDs in astronomy in the first half century of its existence.

The IfA owns and operates the University of Hawai'i's 88-inch (2.2-meter) telescope, seen here circa 1970 when it was the largest telescope on Maunakea. PHOTO: INSTITUTE FOR ASTRONOMY ARCHIVES, UNIVERSITY OF HAWAI'I

University of Hawai'i's 88-inch Telescope

The oldest telescope still in continuous use on Maunakea, the University of Hawai'i's 88-inch (2.2-meter) telescope, had a difficult birth. Construction began in the fall of 1967, but two years of unusually harsh winter weather delayed the telescope's completion until 1970. Getting the telescope's computer control system to work properly—it was new technology at the time—also took several additional years of effort.

Despite those early challenges, the telescope quickly became a scientifically invaluable member of the community of observatories atop Maunakea. It has been a workhorse for the University of Hawai'i for more than four decades, providing observing experience and data for faculty and students alike. Among its many important scientific contributions was the 1992 discovery of QB1, the first object found in the Kuiper Belt, a region of small asteroid-sized bodies and dwarf planets in the outer regions of our solar system.

Today, the 88-inch telescope is usually operated remotely from Hilo or Honolulu. Its protective dome bears the scars of a half century of Maunakea's harsh conditions. PHOTO: MICHAEL WEST

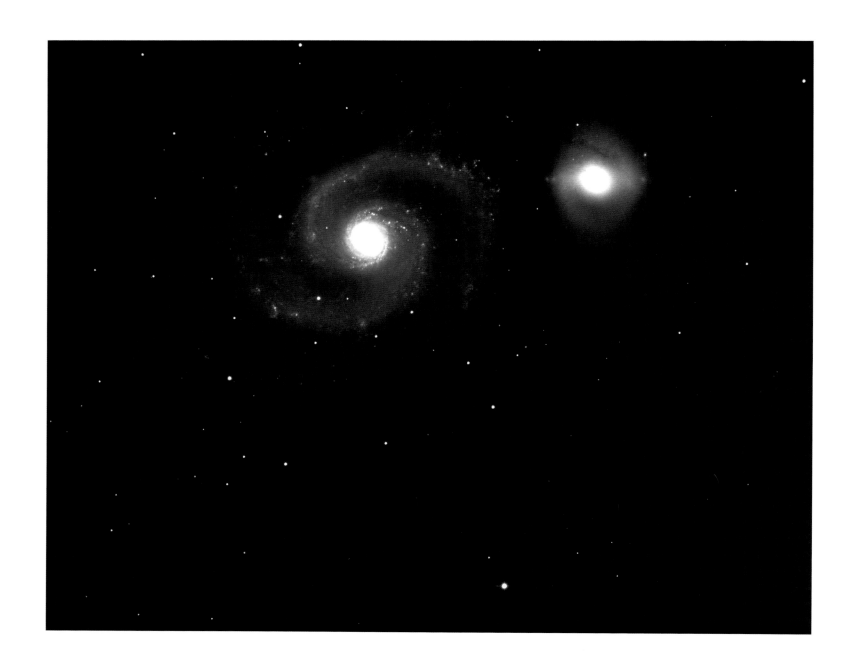

Inventive Instrument Builders

Since its inception, the University of Hawai'i's Institute for Astronomy (IfA) has been a leader in developing new instrumentation for telescopes on Maunakea and beyond, including the Hubble Space Telescope.

A good example is this image of the Whirlpool Galaxy taken by the University of Hawai'i 88-inch (2.2-meter) telescope in 2012, the first obtained using its new sixteen-megapixel infrared sensor. This supersensitive infrared camera, the largest in the world, was the culmination of a twenty-year effort at the IfA that has produced five generations of increasingly larger and more powerful infrared detectors. PHOTO: UNIVERSITY OF HAWAI'I INSTITUTE FOR ASTRONOMY

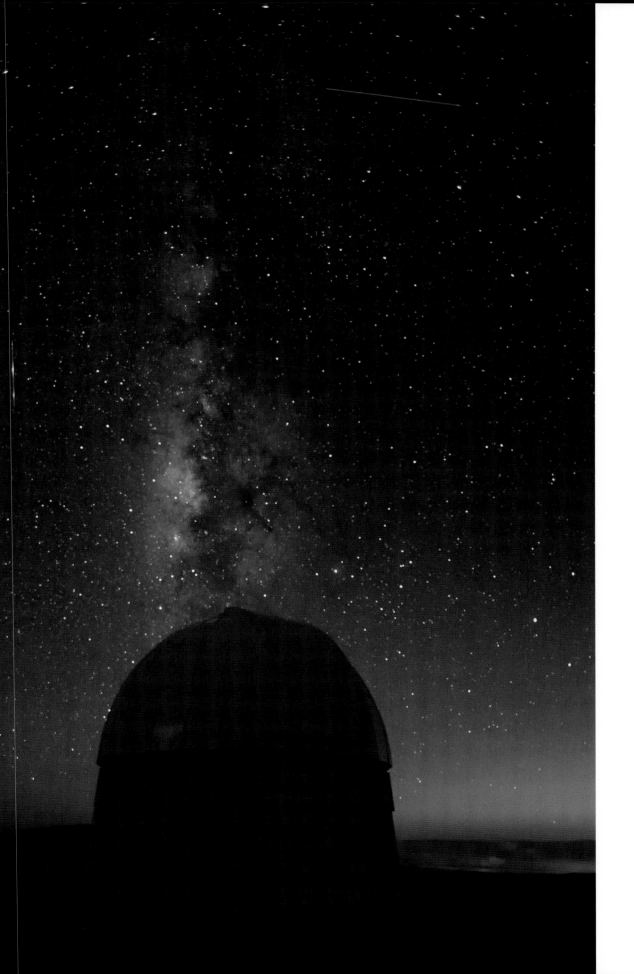

University of Hawai'i at Hilo Educational Telescope

The first research telescope on Maunakea arrived in 1968. Funded by the U.S. Air Force and operated by the University of Hawai'i, this 24-inch telescope was used to track satellites as well as to observe solar system objects such as asteroids and planets. The air force eventually donated the telescope to the University of Hawai'i, where it provided valuable hands-on research experience for students from the Hilo campus for many years.

Four decades of wear and tear eventually took its toll, and in 2008 the original telescope was removed to make room for a larger telescope partly funded by the U.S. National Science Foundation. Named Hōkūke'a, after the Hawaiian name for the Southern Cross, the new observatory has been plagued by technical problems. The University of Hawai'i at Hilo is currently planning to refurbish Hōkūke'a once again with a new state-of-the-art educational telescope that will provide invaluable training opportunities for students interested in pursuing careers in astronomy. PHOTO: MICHAEL WEST

Groundbreaking for the Canada-France-Hawai'i Telescope

Every great construction project begins with a groundbreaking ceremony. This photo shows groundbreaking for the Canada-France-Hawai'i Telescope on July 2nd, 1974. Among the dignitaries were representatives of the three telescope partners: John Jefferies (third from left, director of the University of Hawai'i's Institute for Astronomy), Roger Cayrel (fourth from left, first director of the CFHT), and Ken Wright (second from right, director of Canada's Dominion Astrophysical Observatory). PHOTO: INSTITUTE FOR ASTRONOMY ARCHIVES, UNIVERSITY OF HAWAI'I

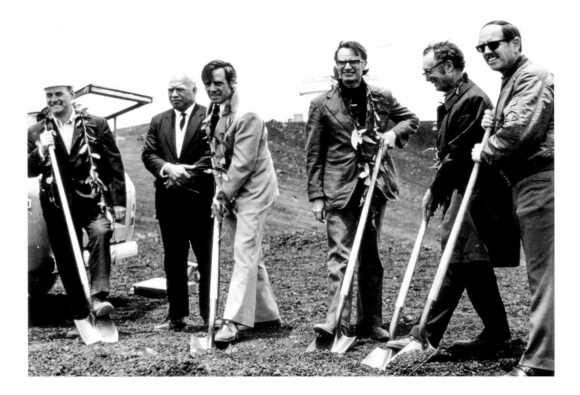

Building an International Partnership

The Canada-France-Hawai'i Telescope (CFHT) was the first international alliance on Maunakea, and its success paved the way for telescopes of eleven nations that reside there today.

Each of the three partners contributed in different ways to CFHT's construction. Canada was responsible for polishing the mirror, building the dome, and developing the telescope's control system. France manufactured and assembled key mechanical components such as the telescope tube seen here. The University of Hawai'i provided a site for the observatory on Maunakea, a road to reach it, and several instruments. In return, Canadian and French astronomers share 85 percent of the available observing time, and University of Hawai'i astronomers receive 15 percent.

CFHT's research capabilities have attracted new partners in recent years, and agreements have been signed with astronomical organizations in Brazil, China, South Korea, and Taiwan. Financial contributions from these additional partners help fund new instruments for CFHT, and astronomers from partner organizations participate in the use of the telescope. The possibility of upgrading the current 11.8-foot (3.6-meter) mirror to create a "next-generation CFHT" known as the Maunakea Spectroscopic Explorer is under consideration, as the observatory is reputed to have one of the best sites on Maunakea's summit ridge. PHOTO: CANADA-FRANCE-HAWAI'I TELESCOPE

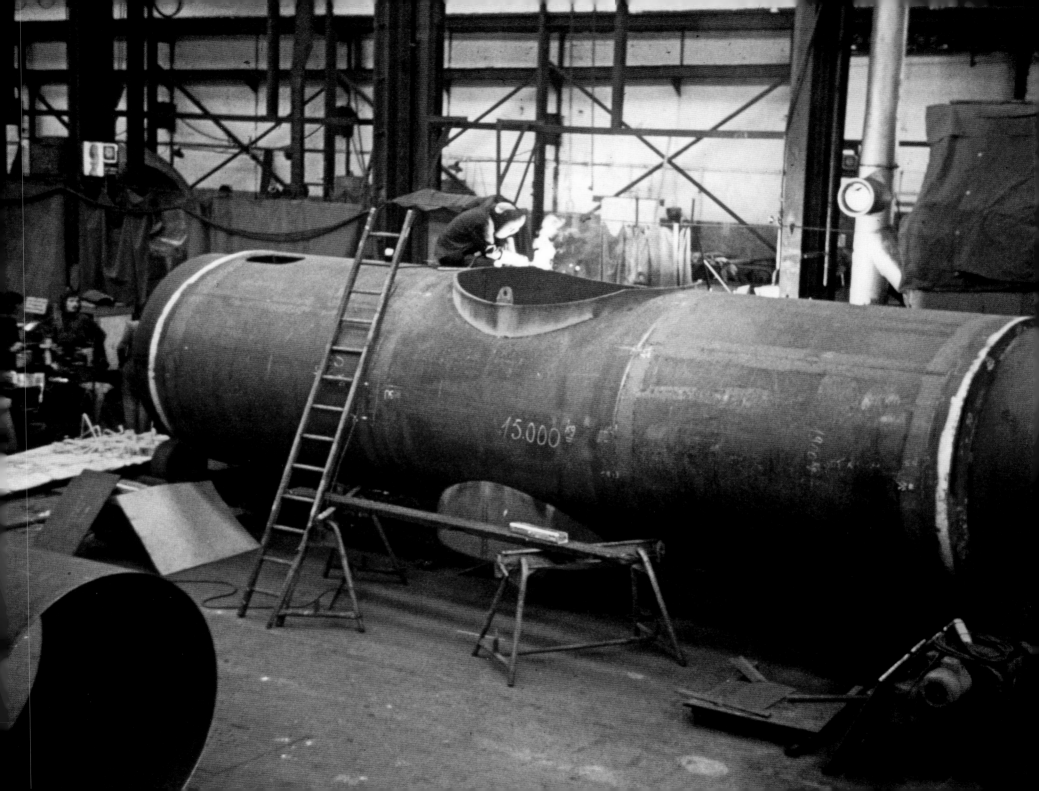

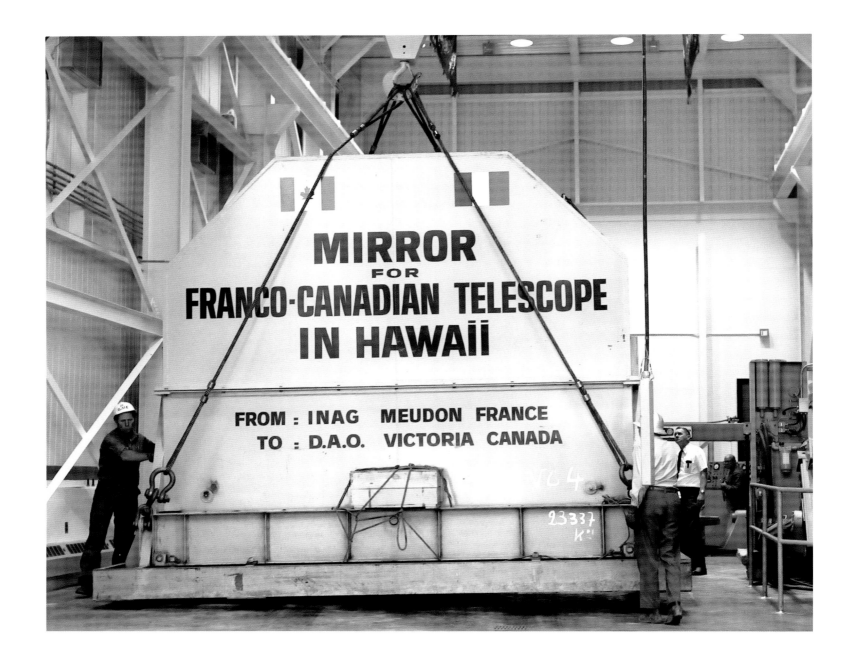

Mirror in Transit

The heart of any telescope is its mirror. Here, the Canada-France-Hawai'i Telescope's mirror, measuring 11.8 feet (3.6 meters) in diameter and tipping the scale at 14 tons, is readied for shipment to the Dominion Astrophysical Observatory in Canada, where it was polished and prepared for installation on Maunakea.

A joint venture between the astronomical communities of Canada, France, and the University of Hawai'i, CFHT has been capturing stunning views of the heavens for nearly four decades. It was the sixth largest optical telescope in the world when it was completed in 1979. PHOTO: CANADA-FRANCE-HAWAI'I TELESCOPE

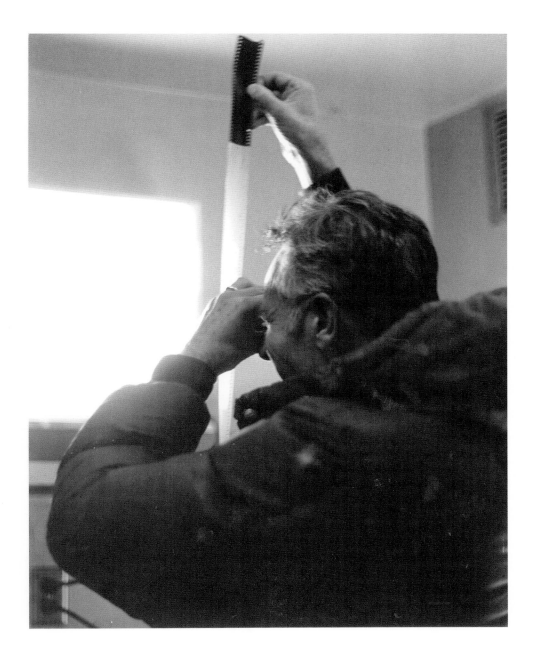

Like a newborn baby opening its eyes for the first time, a new telescope's first glimpse of the heavens—known as "first light"—is an exciting moment that is the culmination of years of effort.

Here, Roger Cayrel, first director of the Canada-France-Hawai'i Telescope, excitedly examines the first image taken with the new CFHT on August 6, 1979. The first images were taken with a simple 35-mm consumer camera because the large-format glass photographic plates favored by astronomers in those days were not readily available in Hawai'i, as they had to be kept cold with dry ice during shipment and storage. PHOTO: CANADA-FRANCE-HAWAI'I TELESCOPE

First Light

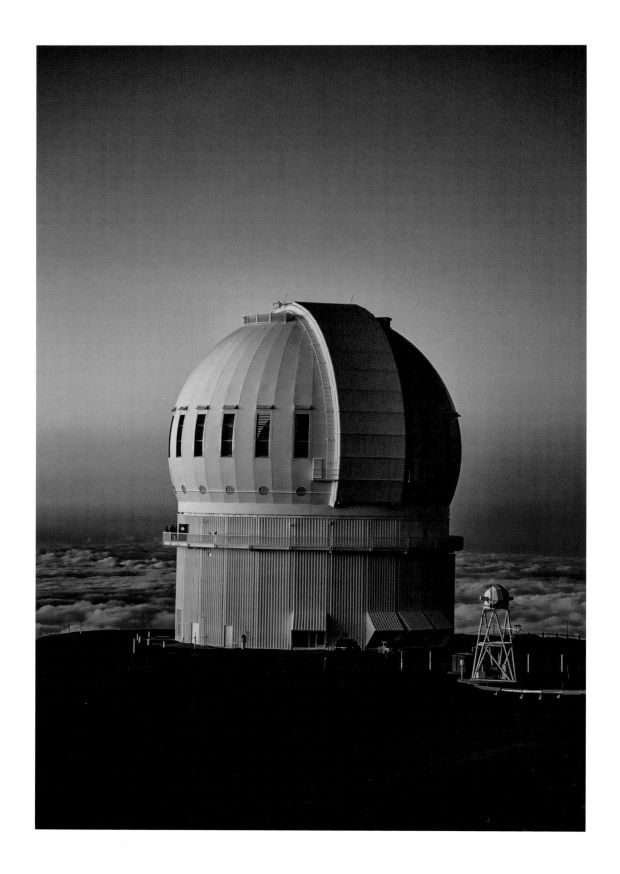

A New Look for the Canada-France-Hawai'i Telescope

Telescopes aren't static fixtures; they're constantly implementing new technologies and new features. In 2013, three decades after it was built, the Canada-France-Hawai'i Telescope's dome was retrofitted with louvers to allow air to circulate more freely, improving CFHT's already excellent image quality.

PHOTO: MICHAEL WEST

The NASA Infrared Telescope Facility

The NASA Infrared Telescope Facility (IRTF) is one of two telescopes on Maunakea dedicated solely to infrared observations, the other being the United Kingdom Infrared Telescope (UKIRT).

The IRTF was built to provide ground-based observations in support of NASA's space missions to explore our solar system. It began operations in 1979 to coincide with the Voyager I spacecraft's flyby of Jupiter. The IRTF made some of the first observations of volcanoes on Jupiter's moon Io, as well as key observations of the size and rotation of Comet Wild 2 before NASA's Stardust spacecraft flew within 149 miles (240 kilometers) of the comet's nucleus in 2004 to scoop up samples of comet dust and return to Earth for study.

The telescope, which is operated by the University of Hawai'i's Institute for Astronomy on behalf of NASA, has a mirror 9.8 feet (3 meters) in diameter. At least 50 percent of the available telescope time is devoted to planetary astronomy and mission support, with the remainder available to the national and international communities for observations of astronomical objects both near and far. Because it collects infrared light, which doesn't get as lost in the sun's glare as visible light does, the IRTF can observe some objects such as comets and planets even during daylight hours. PHOTO: IRTF AND MIKE CONNELLEY

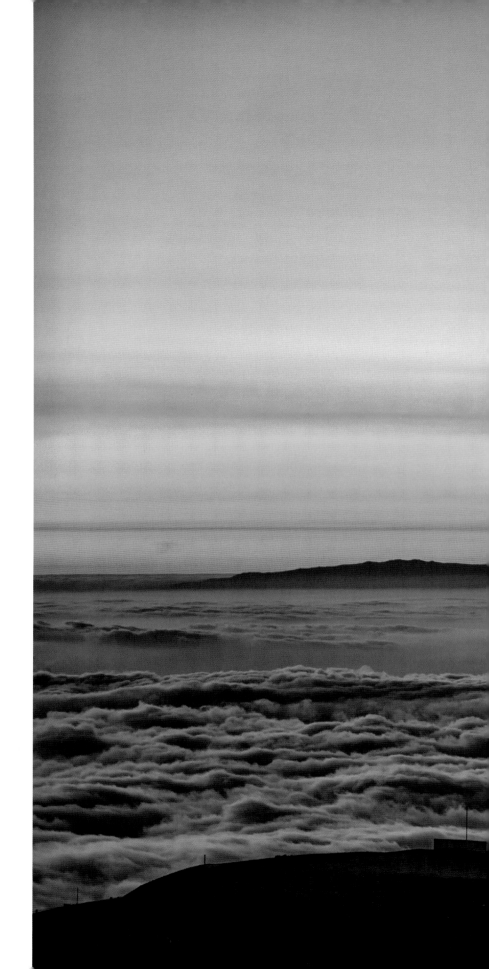

Above the Clouds

Maunakea rises from the depths of the Pacific Ocean as if Earth itself were reaching up to touch the heavens. It's the tallest mountain in the world from base to height, rising 13,796 feet (4,205 meters) above sea level, with another 19,000 feet (5,800 meters) submerged below the ocean's surface. Skies over Maunakea are usually clear thanks to a tropical inversion layer that traps clouds at lower elevations, allowing astronomical observations to be made an average of three hundred nights per year.

Here, NASA's Infrared Telescope Facility (IRTF) is seen above a sea of clouds at sunset. Haleakalā, the tallest peak on the neighboring island of Maui, is visible in the distance. PHOTO: MICHAEL WEST

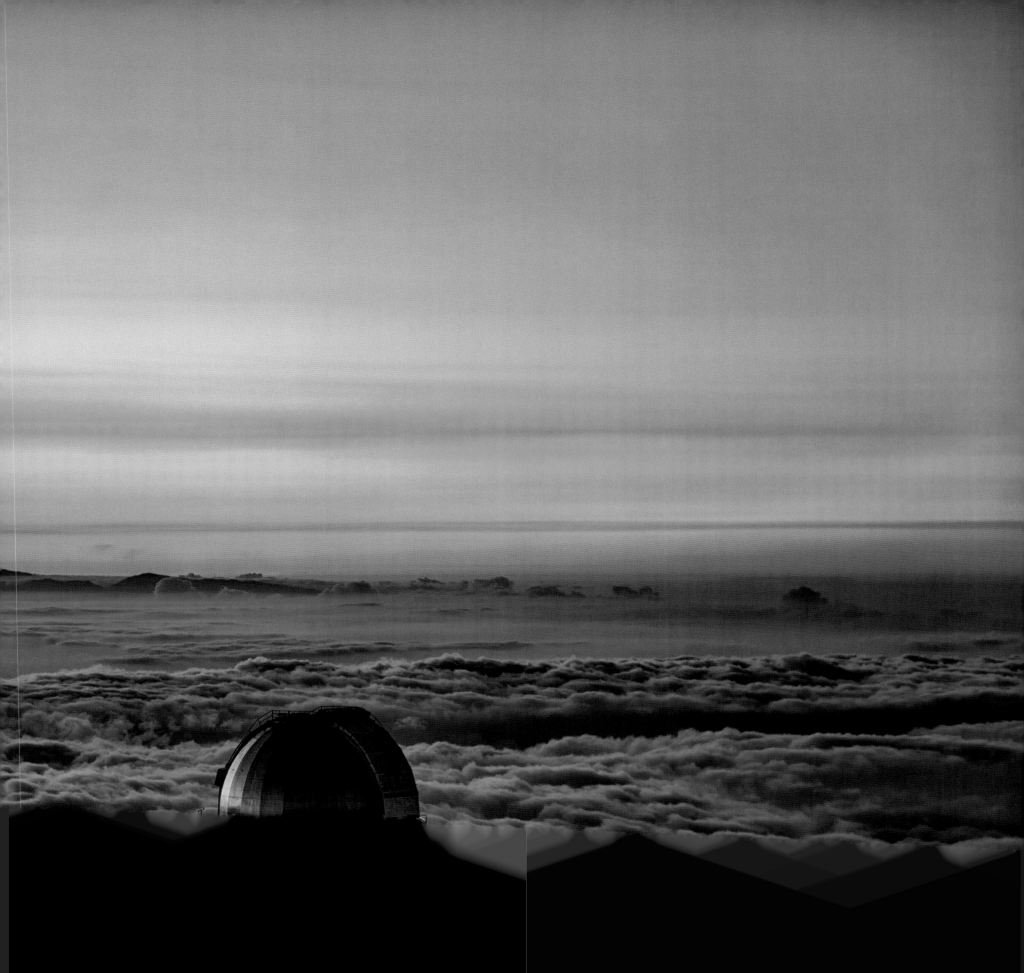

The United Kingdom Infrared Telescope

The United Kingdom Infrared Telescope (UKIRT), owned by the United Kingdom Science and Technology Facilities Council and operated by the Joint Astronomy Centre based in Hilo, saw first light in October 1979. For more than two decades, its 12.5-foot (3.8-meter) mirror made it the largest telescope in the world devoted to infrared observations. Built at a modest cost, UKIRT has been upgraded over the years with cutting-edge instruments that have made it one of the most scientifically productive telescopes on Maunakea.

In 2012, despite its continuing high scientific productivity, the UK government's austerity agenda reached across the world, and the council announced plans to cease funding the operation of UKIRT in September 2013. The observatory issued an Announcement of Opportunity to solicit new operators and, in an ownership transfer process that is unprecedented on Maunakea and perhaps anywhere in the world, UKIRT is now owned by the University of Hawai'i and operated in partnership with the University of Arizona and the Lockheed Martin Advanced Technology Center. In the near term, the telescope will be used primarily to track satellite debris in orbit around our planet and for the study of near-Earth asteroids. PHOTO: MICHAEL WEST

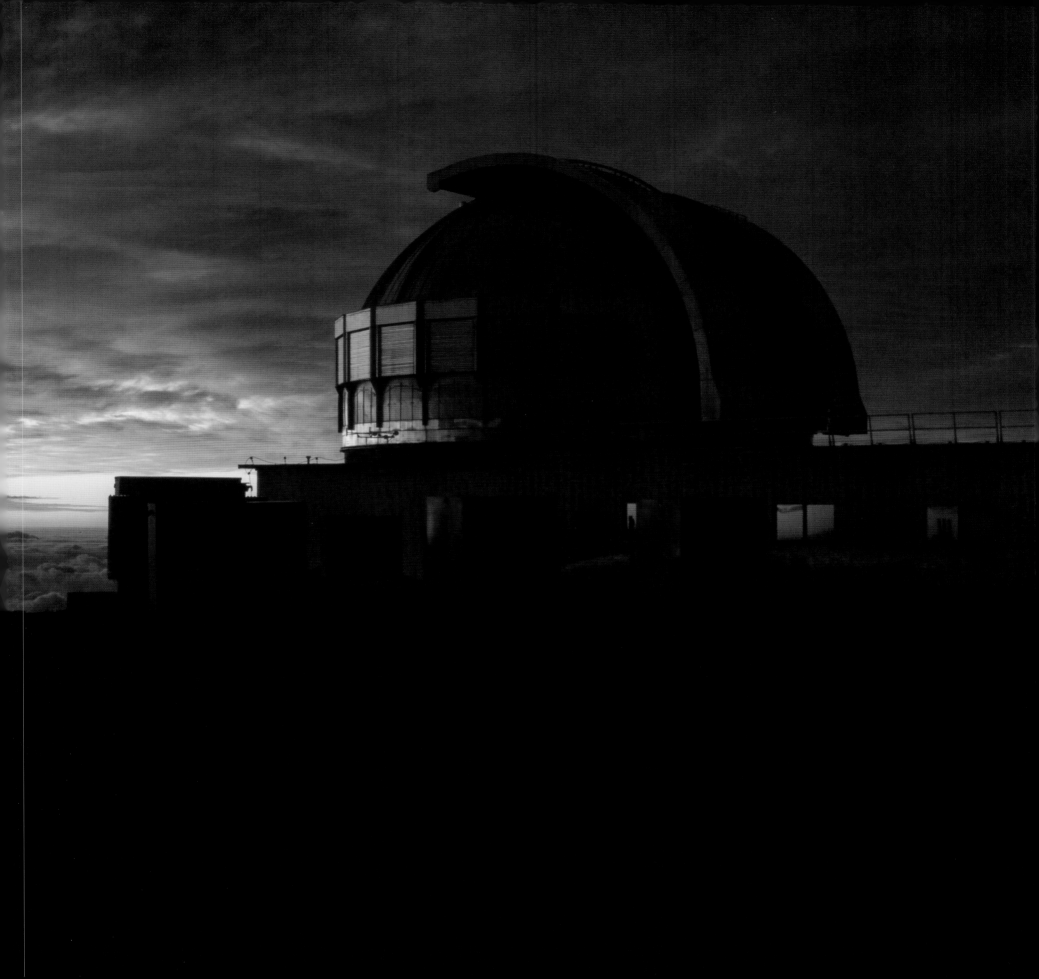

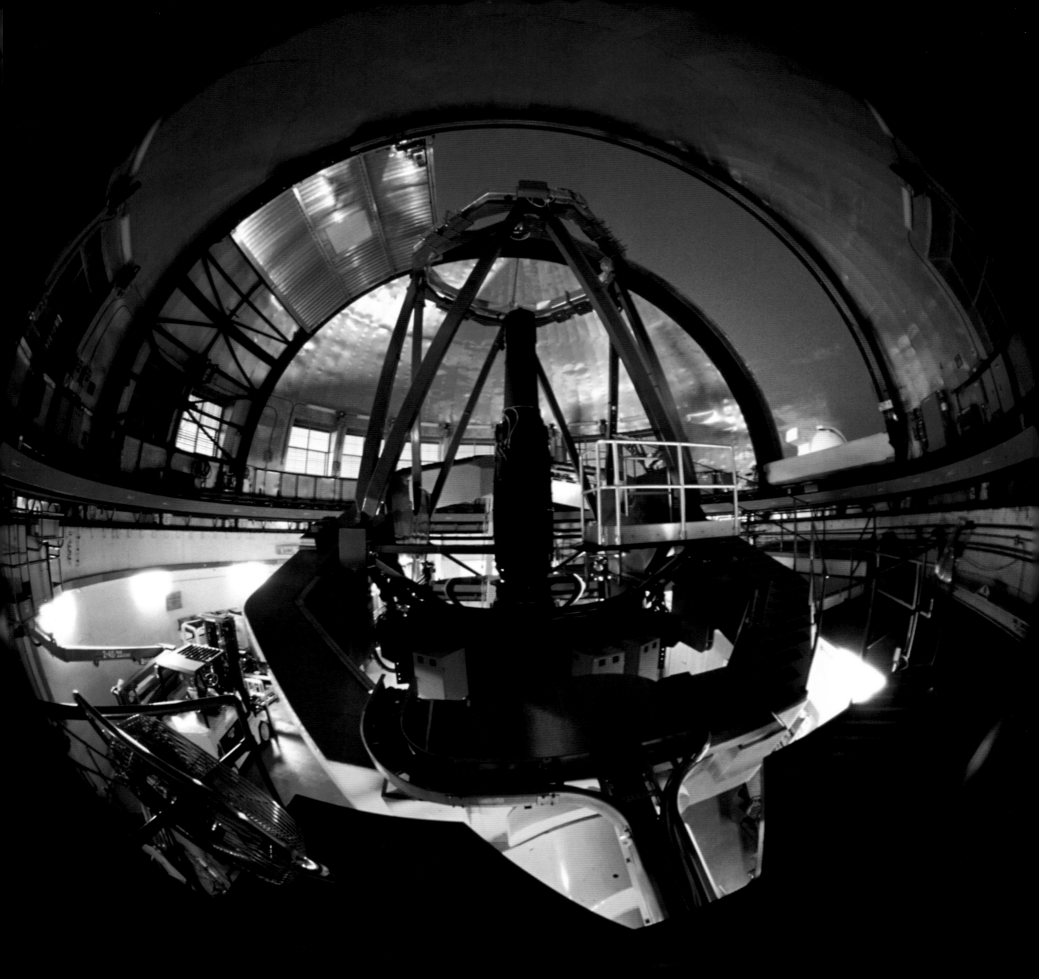

The United Kingdom Infrared Telescope (UKIRT) is one of two telescopes operated by the Joint Astronomy Centre, the other being the James Clerk Maxwell submillimeter telescope. As its name implies, UKIRT captures infrared light that lies just beyond the visible spectrum, providing stunning views of the "warm" universe. This photograph, taken with a fisheye lens, shows the interior of the UKIRT dome.

The black tubular structure immediately above the 12.5-foot (3.8-meter) primary mirror is WFCAM, a wide-field infrared camera designed and built for UKIRT at the UK Astronomy Technology Centre in Edinburgh, Scotland. Between 2005 and 2012, the majority of observing time on UKIRT was devoted to the UKIRT Infrared Deep Sky Survey (UKIDSS). This innovative survey mapped 7,500 square degrees of the northern sky in infrared light, enabling the study of a wide range of astronomical objects from cool and nearby brown dwarfs to extremely distant galaxies and quasars. In a wonderful example of synergy between the observatories on Maunakea, objects initially discovered in the UKIDSS survey have been studied in greater detail using the larger mirrors and greater light-gathering power of the Gemini North, Keck I, Keck II, and Subaru telescopes. PHOTO: PAUL HIRST AND JOINT ASTRONOMY CENTRE

A Blue Night for an Infrared Telescope

A Golden Sunset at UKIRT

Although the sun's intense glare makes it impossible to see visible light from planets, stars, and galaxies during daylight hours, infrared telescopes can observe for a while before sunset and after sunrise because the sun emits less infrared light, extending the "night" by a few hours for telescopes like the United Kingdom Infrared Telescope. PHOTO: TOM KERR AND JOINT ASTRONOMY CENTRE

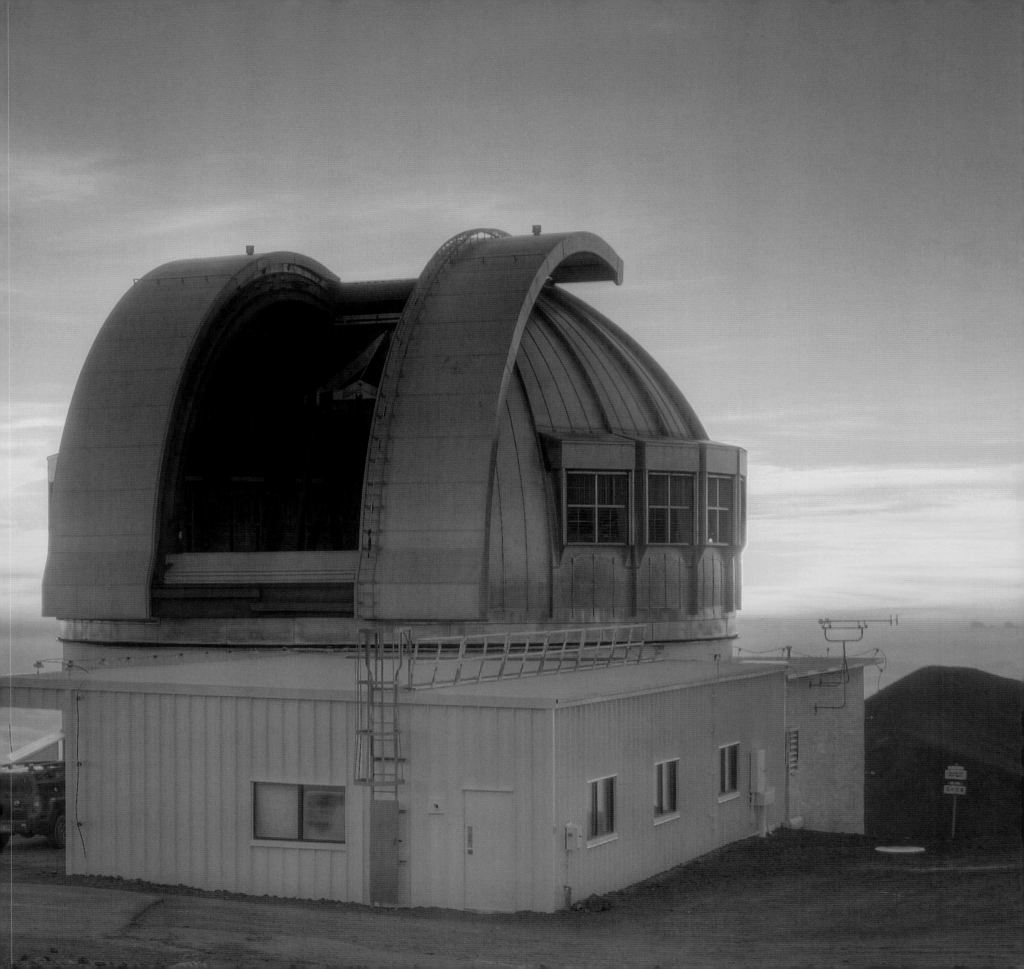

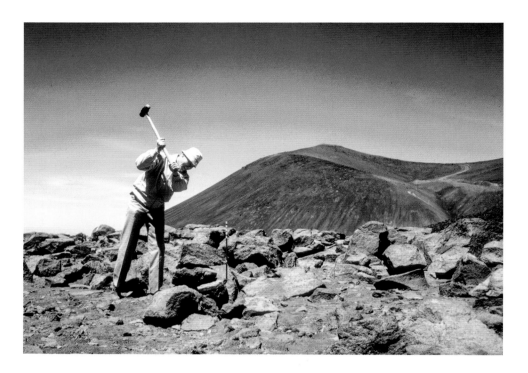

Astronomer Robert Leighton marks his chosen spot for the site of the Caltech Submillimeter Observatory (CSO) near Maunakea's summit in 1979.

The first telescopes on Maunakea harvested only the visible light that falls from the sky. One of the most significant changes over the past half century has been the construction of telescopes that collect light our eyes can't see, such as submillimeter light. This particular type of light comes primarily from very cold, dusty regions of the universe, providing new insights into the birth of stars and planets, the evolution of galaxies, cosmology, and a host of other areas of astronomical research.

The CSO was designed by Leighton and built at Caltech. The telescope's mirror is made of eighty-four lightweight aluminum panels that together span a diameter of 34 feet (10.4 meters). The CSO is testament to Leighton's vision of building a submillimeter telescope on the best site in the world.

After two decades of service, in 2009 Caltech announced plans to shut down the CSO to focus its resources and research efforts on a next-generation submillimeter telescope in Chile. Currently it is expected that the CSO will be dismantled in 2016, and the site where it stood on Maunakea will be returned to its original pristine state by 2018. PHOTO: CSO

A Dream Takes Shape

The Caltech Submillimeter Observatory at Night

Internal lights bathe the Caltech Submillimeter Observatory (CSO) in a green and red glow at night. CSO is the only telescope on Maunakea that doesn't have a telescope operator to control the telescope. Instead, astronomers do everything during the night, controlling the telescope and the instrument that collects data. PHOTO: JASON CHU

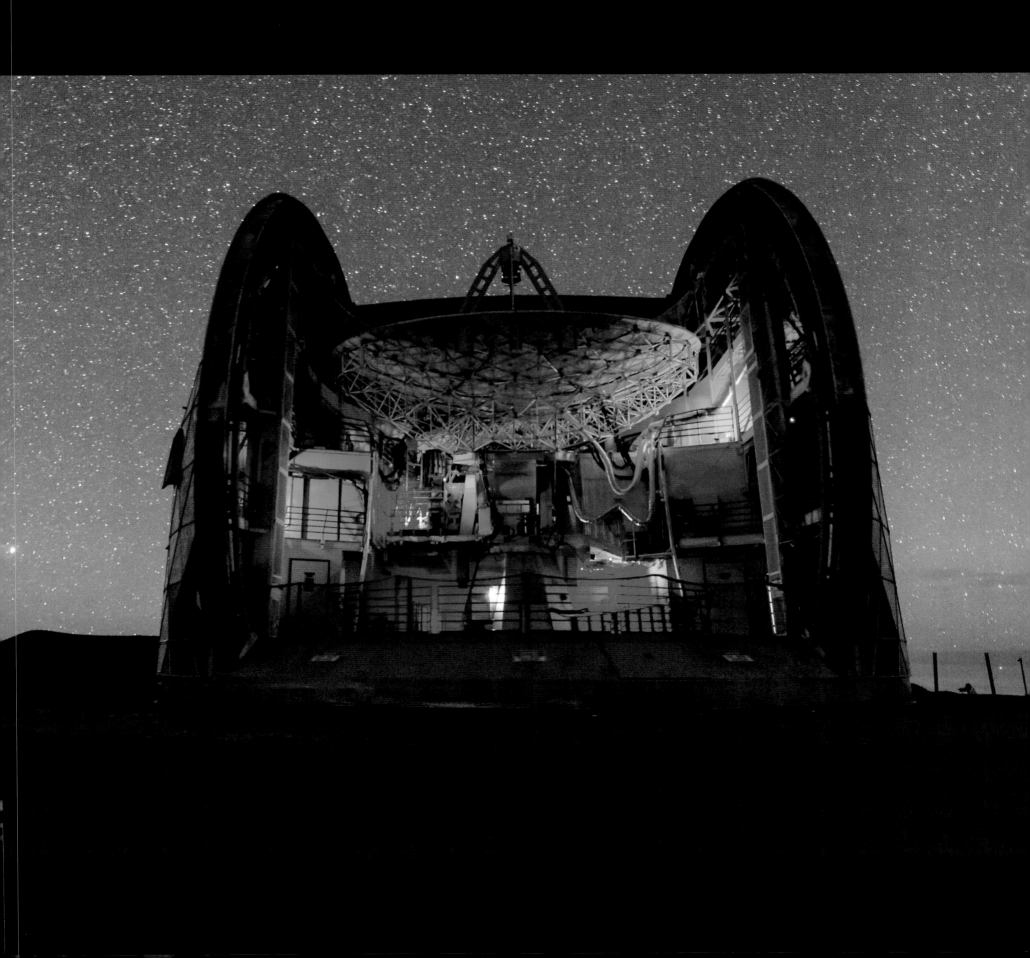

Catching
Submillimeter
Light

Submillimeter telescopes have similarities to both infrared and radio telescopes, collecting microwaves that have less energy than infrared light but more energy than radio waves. Submillimeter light is especially valuable for learning about very cold and dusty regions of space—the stuff planets and stars are made from.

The Caltech Submillimeter Telescope (CSO) looks like a radio telescope, with its 34-foot (10.4-meter) metallic dish encased within a protective dome like an infrared telescope. The dish itself is composed of interlocking hexagonal segments made of lightweight aluminum, a design strategy that would also be used with great success by Keck Observatory. PHOTO: MICHAEL WEST

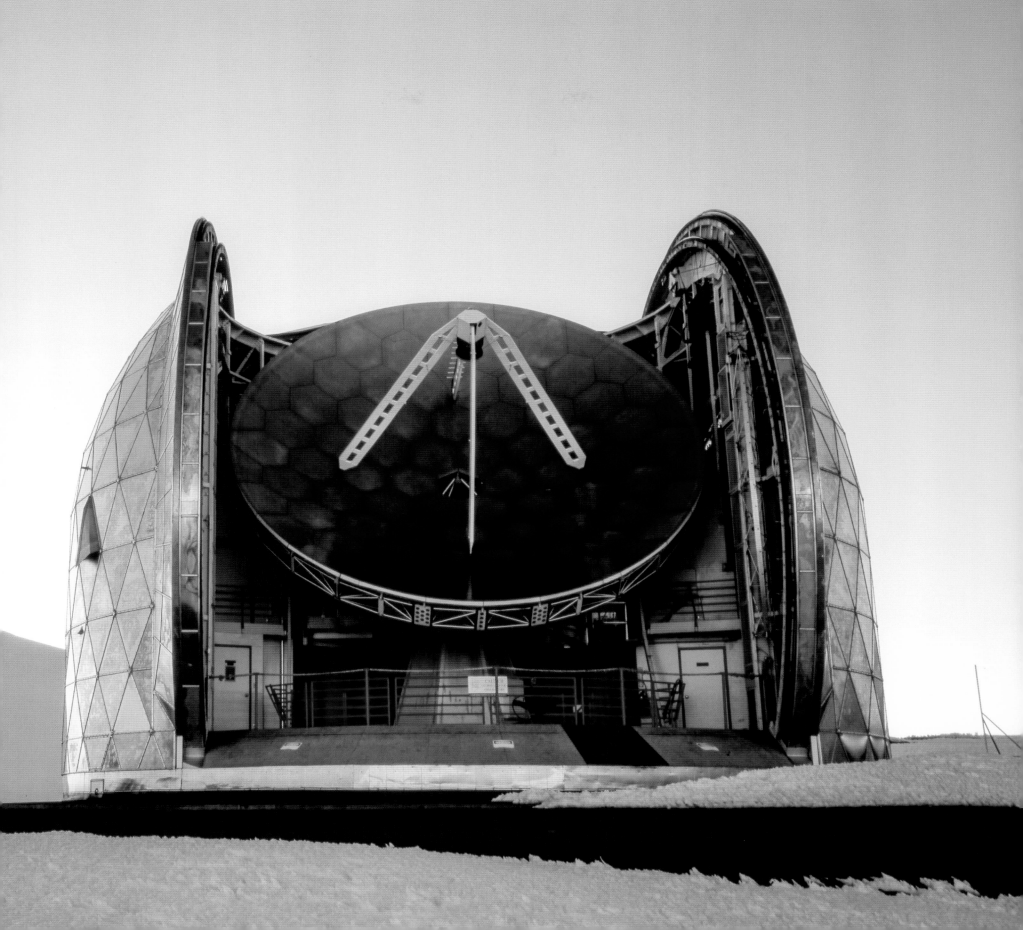

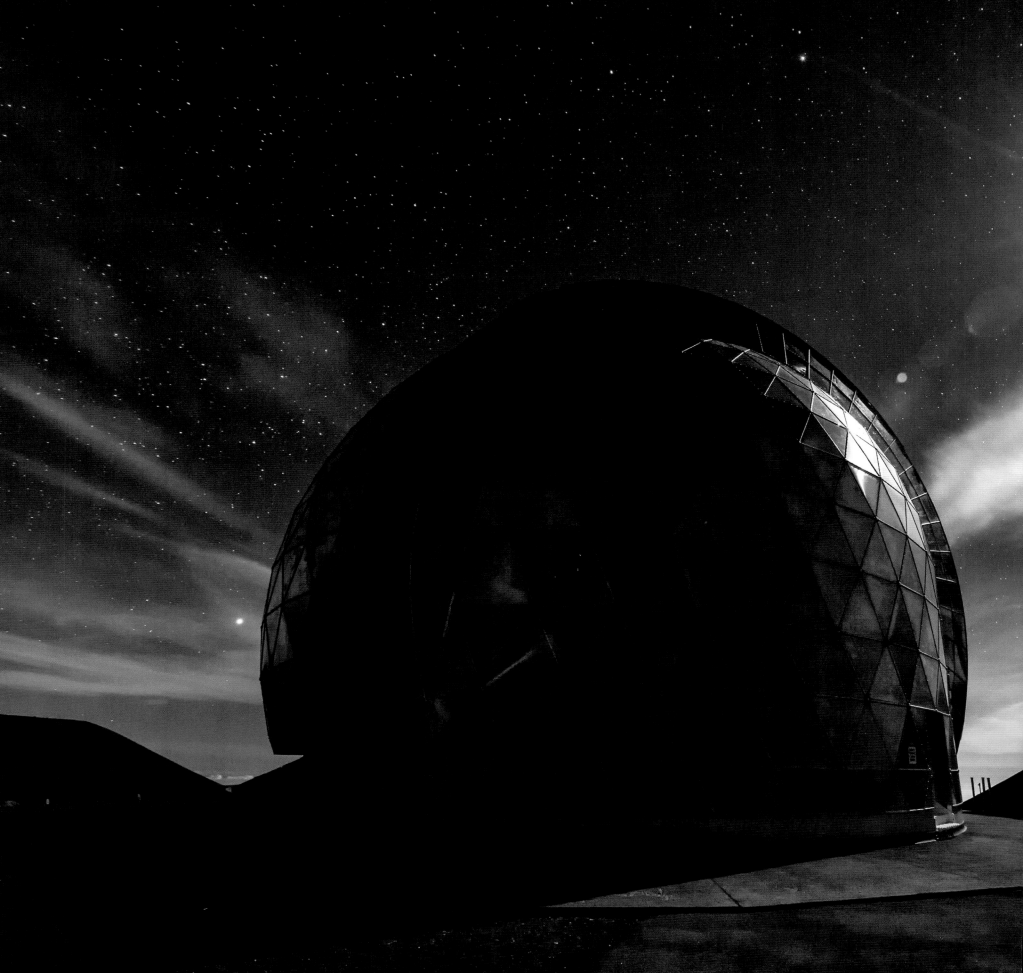

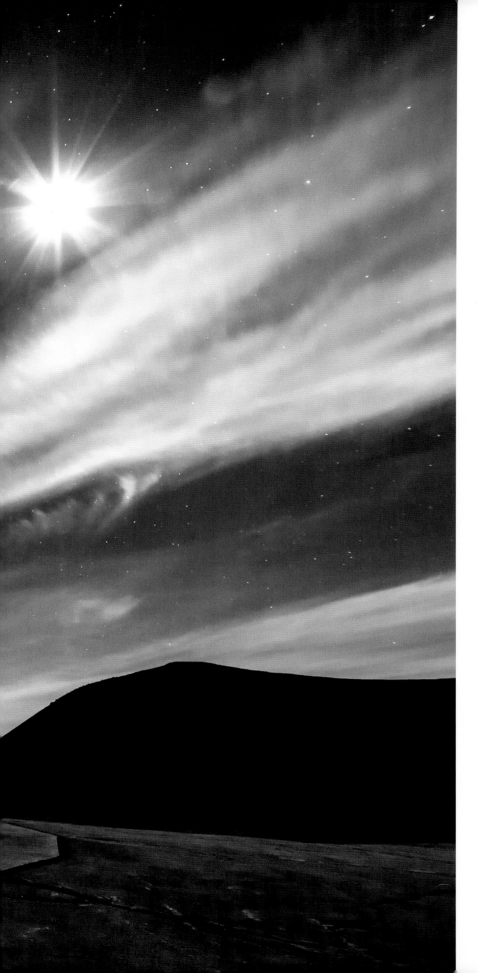

Caltech Submillimeter Observatory

The first submillimeter telescope on Maunakea, the Caltech Submillimeter Observatory (CSO) has been observing the universe since 1986. Its futuristic-looking metallic dome houses a metal dish 34 feet (10.4 meters) in diameter that is used to collect microwaves—submillimeter light—from cold and dusty regions of space, the kinds of places where planets and stars are born. PHOTO: MICHAEL WEST

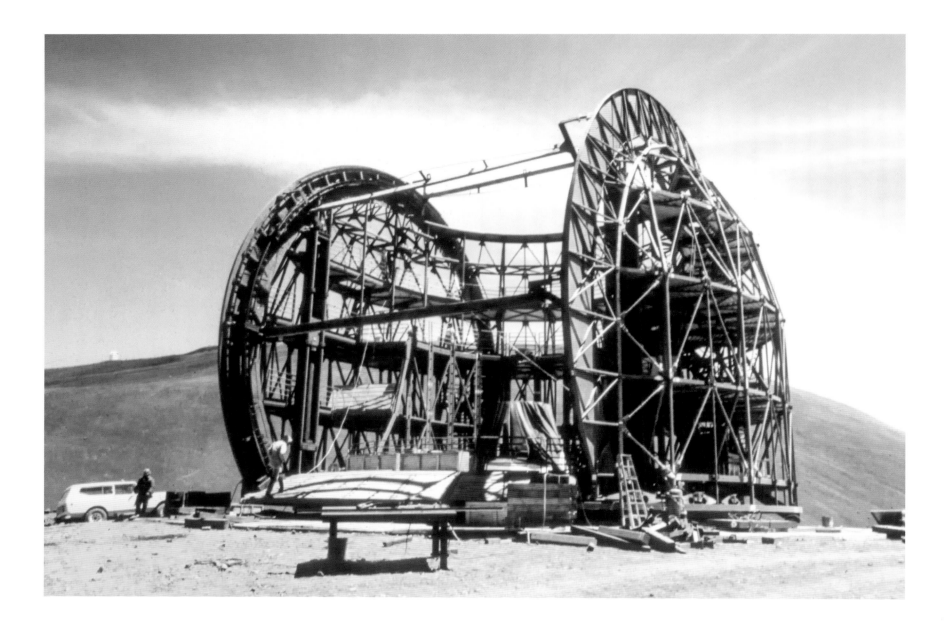

Constructing
the Caltech
Submillimeter
Observatory

The first submillimeter telescope on Maunakea, the Caltech Submillimeter Observatory (CSO) began observing the cosmos in 1986. Every component of the telescope, including its protective dome, was built and tested at Caltech in Pasadena, California, before being shipped to Hawai'i. Here the partially assembled CSO dome is seen on Maunakea in 1986. PHOTO: CSO

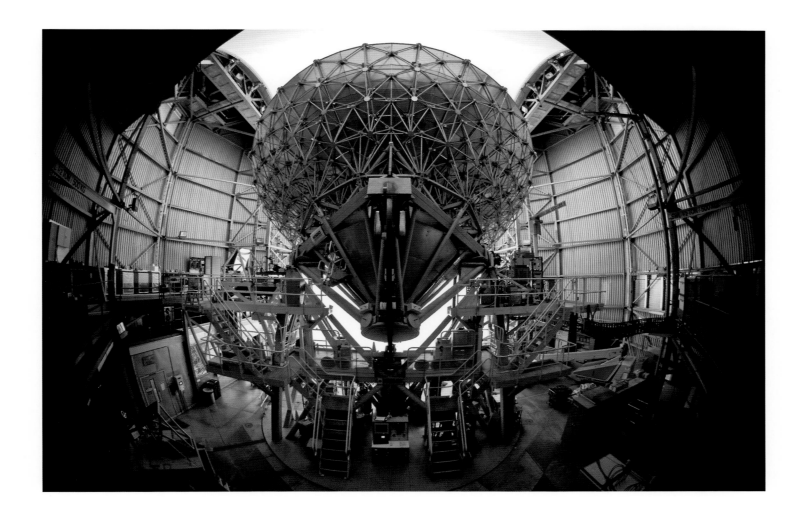

The Irish writer Jonathan Swift once said, "Vision is the art of seeing things invisible."

Stars, galaxies, and other celestial objects emit a rainbow of colors that our eyes can't see. If astronomers studied only the visible light that comes from space, they would be missing much of the picture. Some of the most interesting objects in the universe shine most brightly in light that is invisible to our eyes, such as X-ray, radio, ultraviolet, infrared, and submillimeter light.

To make the invisible universe visible, astronomers use specially designed telescopes and instruments that can record these types of light. For example, the Very Long Baseline Array telescope located on Maunakea uses a large metal dish that acts like a mirror to gather radio waves from space and convert them into a "radio image" of an object. Similarly, the United Kingdom Infrared Telescope, also located on Maunakea, is designed to collect infrared light that lies just beyond the range of what our eyes can see.

This wide-angle photograph shows the interior of the James Clerk Maxwell Telescope. The JCMT collects submillimeter light using a reflective dish that is 49 feet (15 meters) across and weighs 70 tons. It is made from 276 aluminum panels that are held in place by the metallic structure seen in this photo. Each panel can be tilted individually to maintain the mirror's perfect shape. PHOTO: ANTONIO CHRYSOSTOMOU AND JOINT ASTRONOMY CENTRE

Seeing the
Invisible

The Glow of the James Clerk Maxwell Telescope

Named in honor of the nineteenth-century Scottish scientist who first correctly explained the nature of light, the James Clerk Maxwell Telescope (JCMT) is the largest single submillimeter telescope in the world. It was built in 1987 as a partnership between the United Kingdom, Canada, and the Netherlands.

To protect the JCMT from wind, rain, dust, and direct sunlight that could damage its sensitive instruments, the opening of the telescope's dome is shielded by one of the world's largest pieces of Gore-Tex, a waterproof fabric that's also used to make raincoats. Just as a glass window provides protection from the outside elements while still allowing light to pass through, the Gore-Tex that protects the JCMT is transparent to submillimeter light.

The JCMT partnership dissolved in 2013–2014, as all three countries redirected funds to support the new Atacama Large Millimeter/submillimeter Array (ALMA), which is composed of sixty-six moveable radio dishes in the north of Chile. The JCMT is now owned by the University of Hawai'i and operated by the East Asian Observatory in partnership with universities in the United Kingdom and Canada. The East Asian Observatory is supported by research institutes in China, Japan, South Korea, and Taiwan, marking the first time these countries have acted in concert in observational astronomy.

PHOTO: MICHAEL WEST

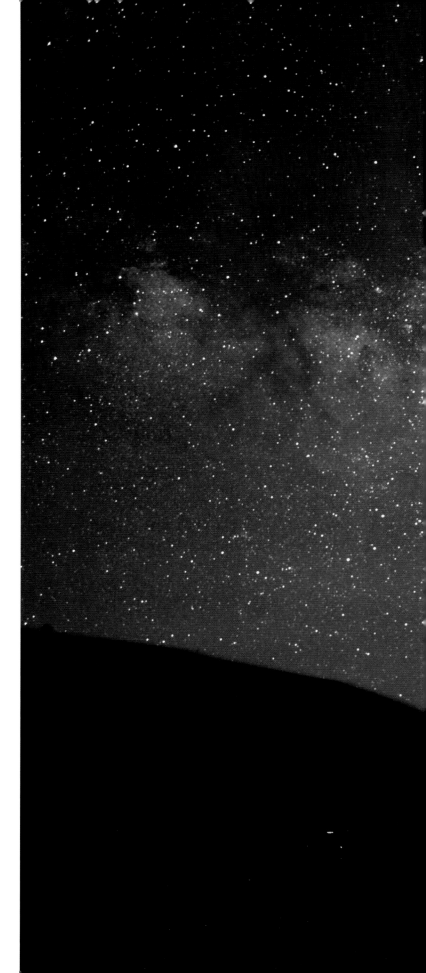

The Milky Way and the James Clerk Maxwell Telescope

It's estimated that two-thirds of Americans can no longer see the Milky Way at night because of the glare of artificial lights where they live. In many cities and towns, only a few of the brightest stars remain visible—the rest have been lost in a luminous fog.

Thanks to the Big Island's low population density—an area almost as big as Connecticut but with only 5 percent its population—as well as lighting ordinances intended to preserve the island's dark skies, skies over Maunakea remain unspoiled by light pollution. Here the Milky Way blazes over the James Clerk Maxwell Telescope. On dark, moonless nights, the light of the Milky Way is bright enough to cast faint shadows across the landscape. PHOTO: MARCEL CLEMENS

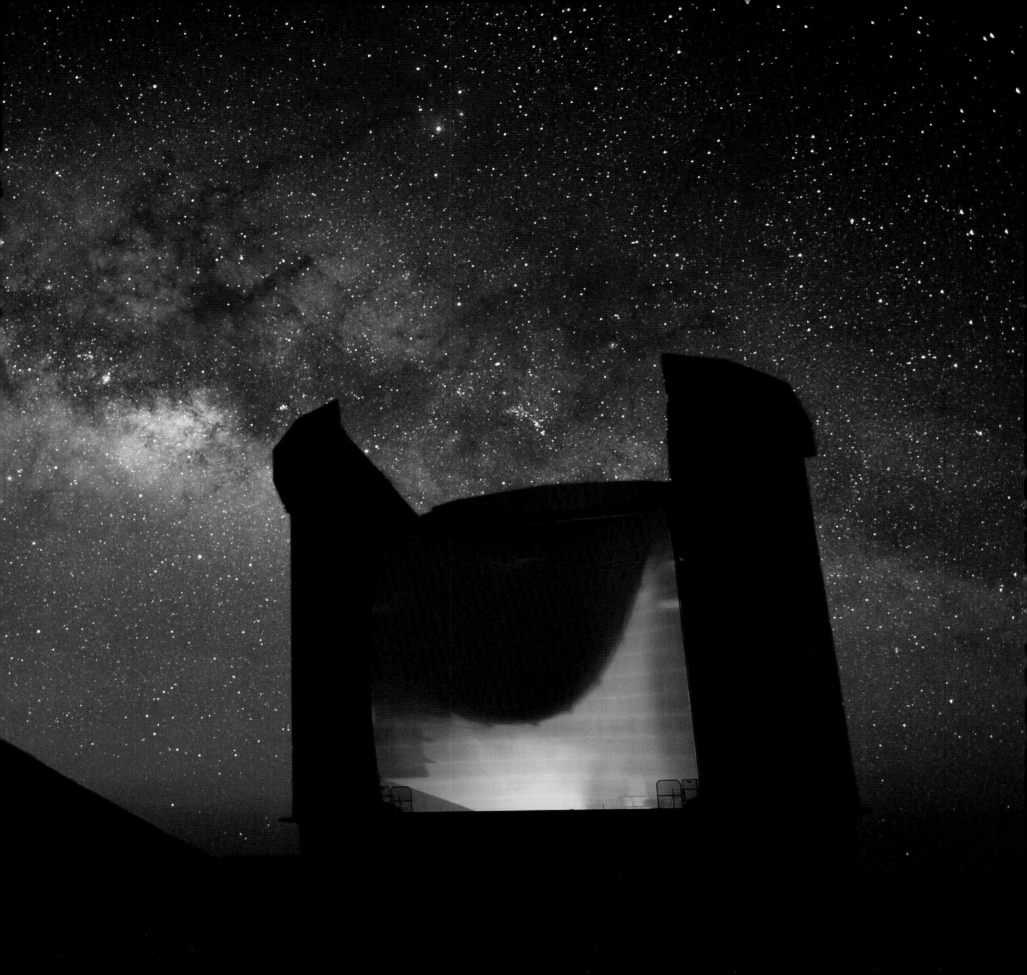

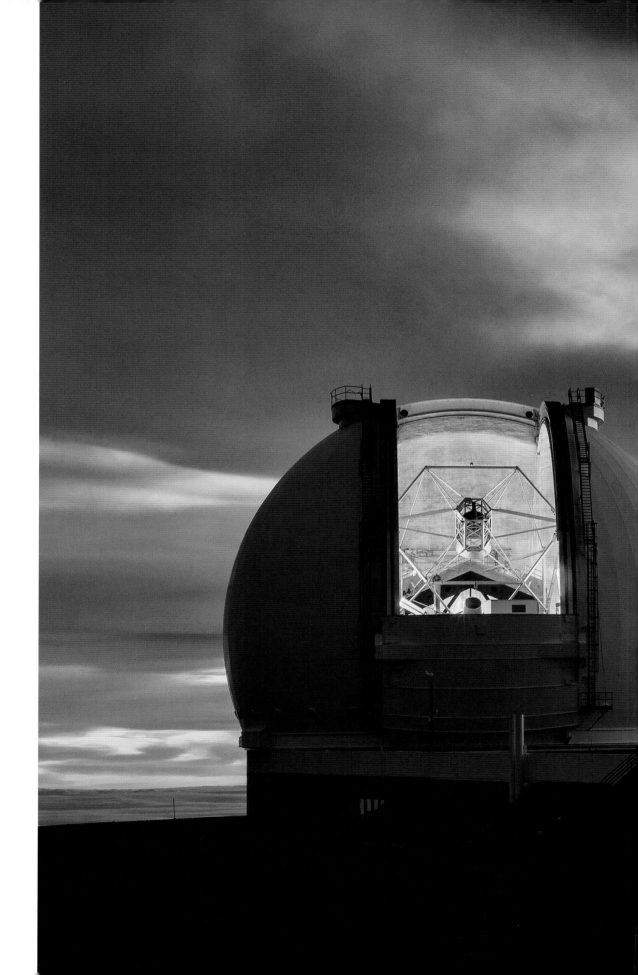

Two Telescopes, One Vision

Maunakea is home to the two largest optical and infrared telescopes in the world: the twin Keck Observatory 33-foot (10-meter) telescopes. Built with a generous gift of $140 million from the W. M. Keck Foundation, the Keck I telescope saw first light in 1992 and its twin, Keck II, followed in 1996. The W. M. Keck Observatory manages the observatory on behalf of the California Institute of Technology, the ten-campus University of California system, and NASA.

Each Keck Observatory telescope stands eight stories tall and weighs 300 tons. This photograph shows the domes as they open to begin another successful night of exploring the universe. PHOTO: W. M. KECK OBSERVATORY AND ETHAN TWEEDIE

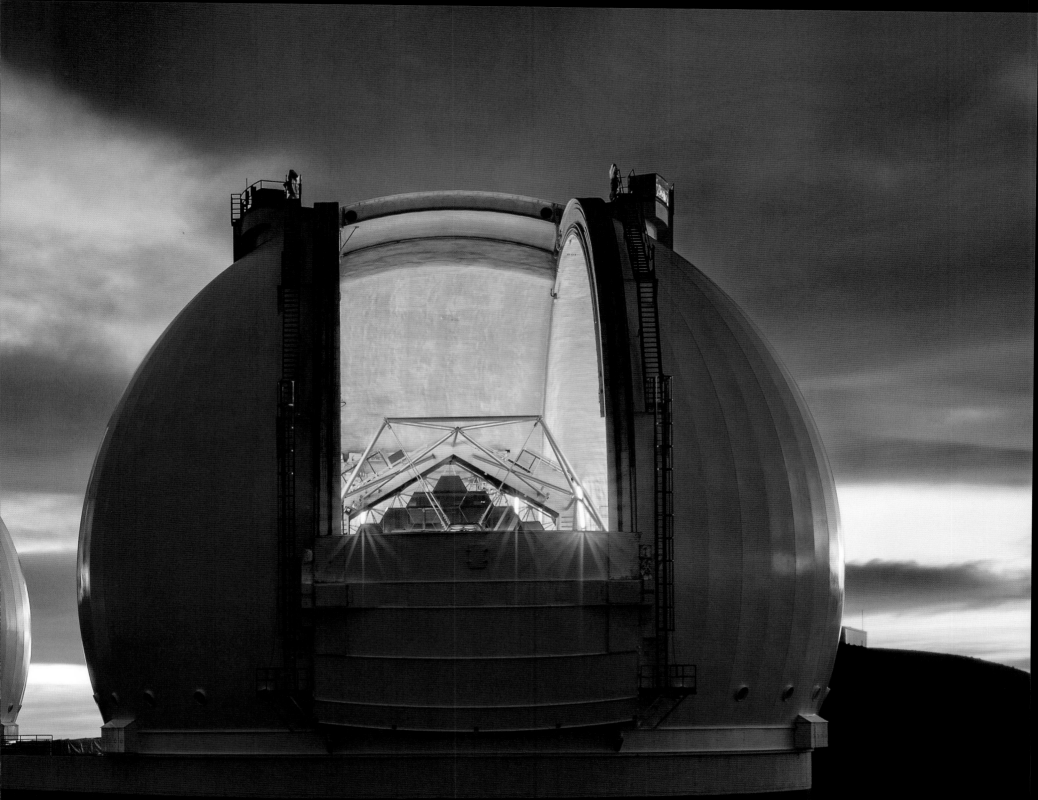

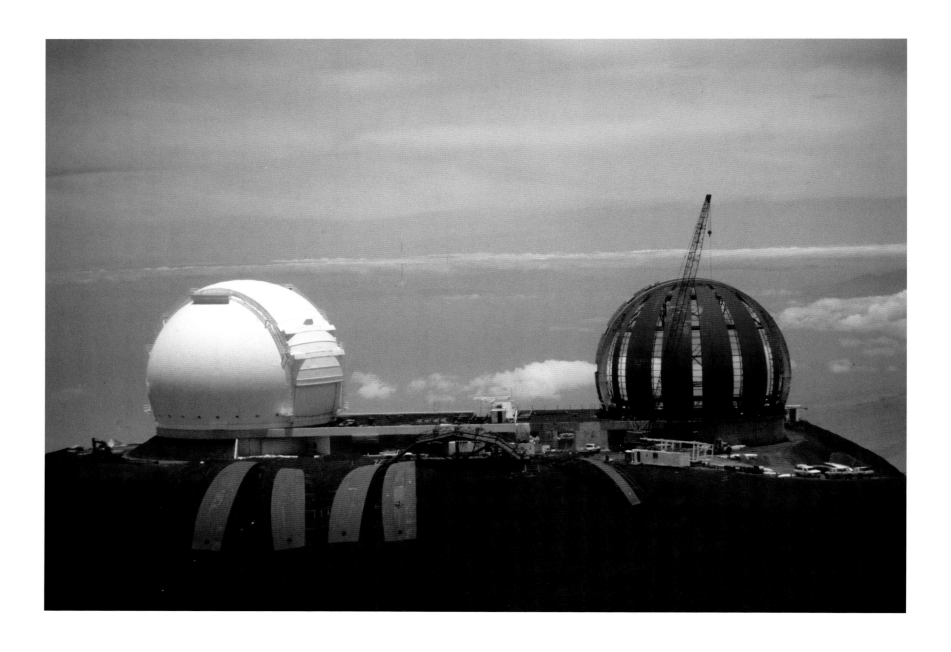

Birth of a Sibling

Inauguration of the Keck I telescope took place on November 7, 1991, and was followed immediately by a groundbreaking ceremony for Keck II. This 1993 photo shows the ongoing construction of Keck II's dome, which was built while its sibling was already in full operation.

Discoveries made with the Keck Observatory telescopes have revolutionized our understanding of the universe for more than two decades. Highlights include a plethora of planets found around other stars, evidence of a supermassive black hole in the Milky Way's center, the discovery of other Pluto-sized objects in the outer solar system, and observations that revealed the accelerating expansion of the universe, research that led to the 2011 Nobel Prize in Physics. PHOTO: W. M. KECK OBSERVATORY

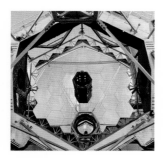

A Revolutionary Design

The heart of every telescope is a curved mirror that collects droplets of light raining down on Earth from the cosmos. Just as a bigger bucket collects more raindrops than a smaller one, a bigger telescope mirror collects more light, yielding brighter and sharper images. Gathering enough light to see the faint flickers of stars and galaxies at the edge of the observable universe requires the largest possible mirrors.

Optical and infrared telescopes use mirrors made of glass coated with a thin layer of reflective material. If the mirror is too big, however, the glass begins to sag under its own weight. Glass also expands and contracts as temperatures change, altering the mirror's shape as the night air cools and degrading the quality of images. For decades, these undesirable properties of glass limited the size of the largest telescope mirrors to less than about 16 feet (5 meters) in diameter.

A radical new design allowed Keck Observatory to build the world's biggest mirrors, each 33 feet (10 meters) in diameter. Instead of a single monolithic piece of glass, each mirror in Keck I and Keck II is composed of thirty-six smaller mirror segments that fit together like pieces of a puzzle. Computer-controlled supports beneath the segments keep them positioned within 1/4,000th of the width of a human hair to maintain the mirror's shape, checking and realigning the segments twice per second if needed.

The Keck I telescope had only nine mirror segments in place when it saw first light in November 1991, yet the quality of the images it produced were already phenomenal. The remaining segments were added one by one over the following months, and by April 1992 the mirror was complete. Keck II followed a few years later, opening its eye on the sky for the first time in January 1996.

Keck Observatory's segmented mirror design has proven so successful that the next generation of extremely large telescopes, including the Thirty Meter Telescope planned for Maunakea, will be based on a similar design. PHOTO: ETHAN TWEEDIE AND W. M. KECK OBSERVATORY

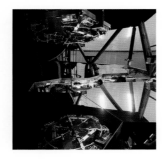

Mirror Segments

Every optical and infrared telescope on Maunakea must have its mirror cleaned and recoated from time to time. For most, this requires removing the entire mirror, often weighing many tons, from the telescope. However, the segmented design of the Keck Observatory's mirrors makes it possible to remove a single segment at a time and replace it with a spare while the original is being cleaned or recoated. Here, Keck Observatory technicians remove one of the hexagonal-shaped mirror segments. PHOTO: W. M. KECK OBSERVATORY AND MARK DEVENOT

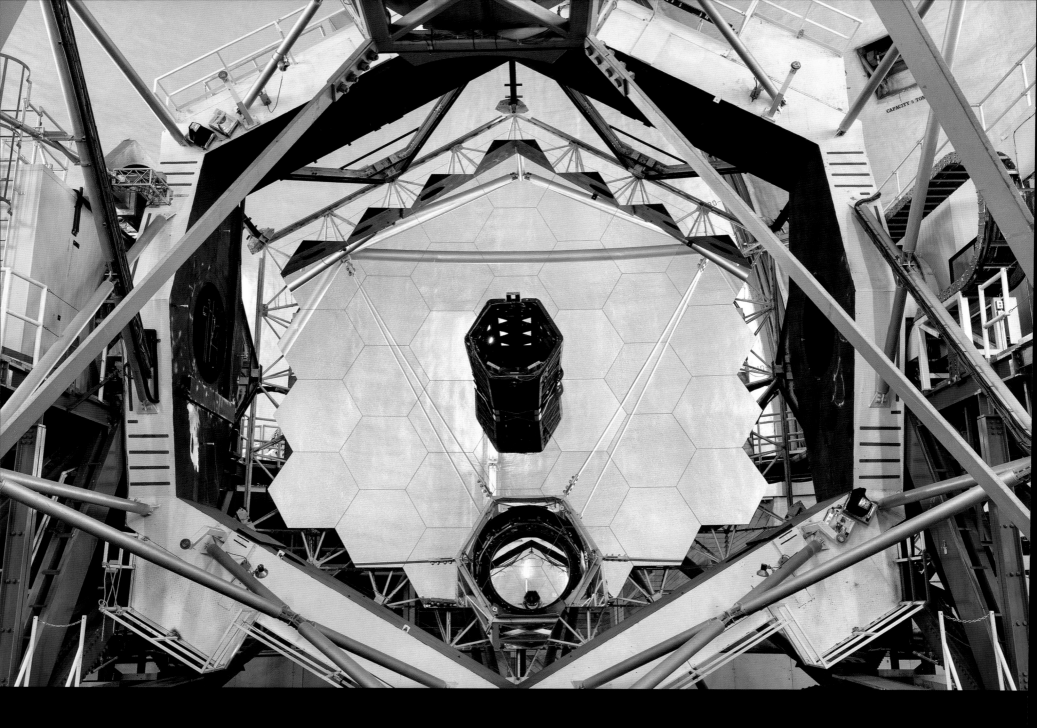

A Revolutionary Design

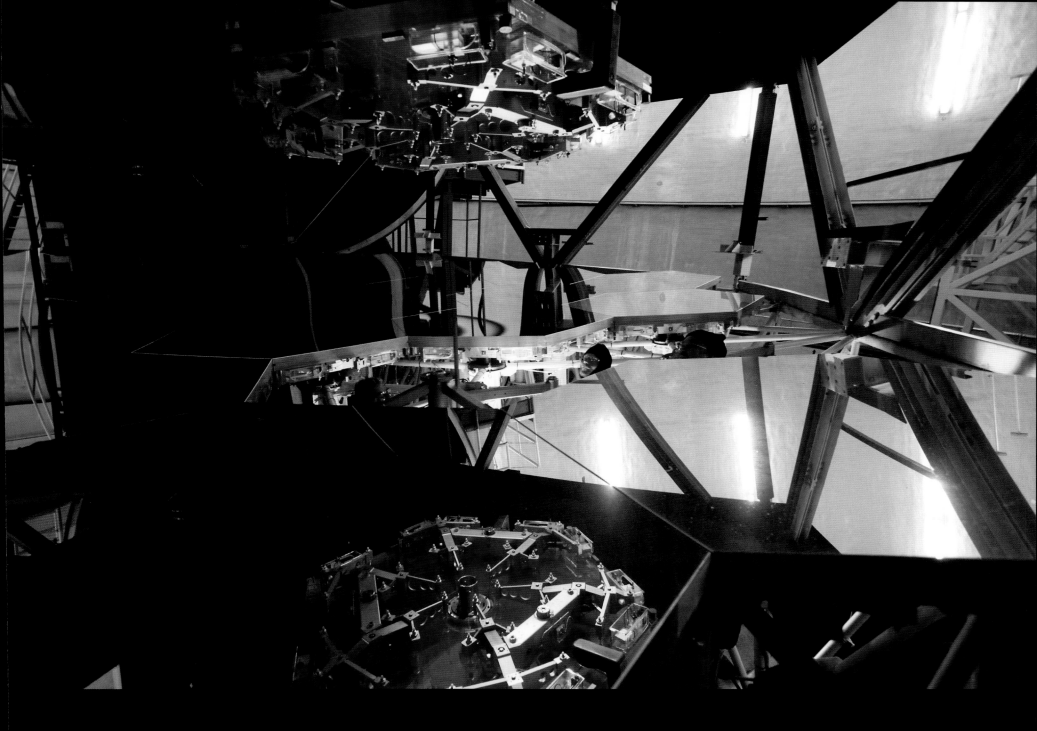

Mirror Segments

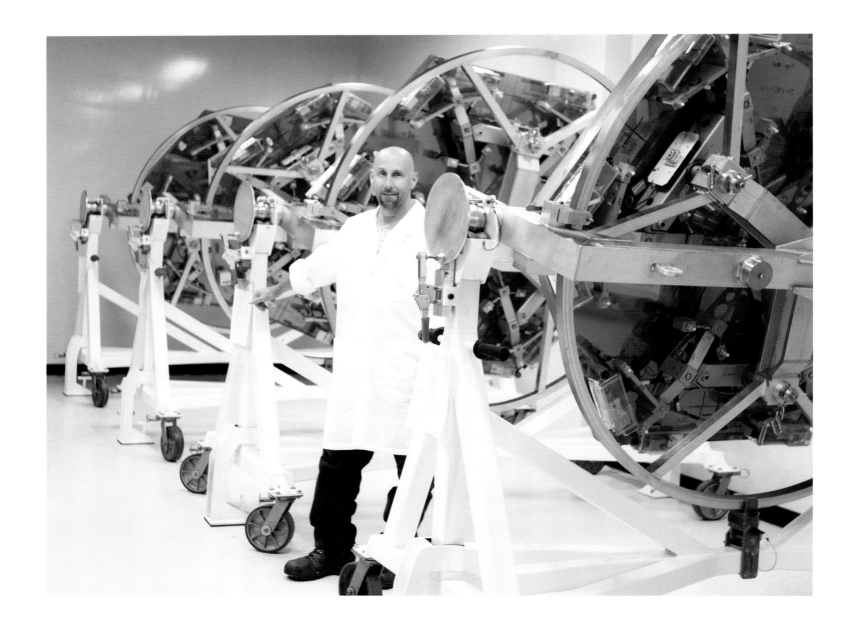

A Modular Mirror

Each of the two Keck Observatory telescopes' mirrors is composed of thirty-six smaller segments that fit together to create the equivalent collecting area of a single mirror 33 feet (10 meters) in diameter. Individual segments can be swapped in and out as needed for cleaning, recoating, or repairs. Each segment is 6 feet (1.8 meters) across and weighs 880 pounds. Optics technician George Wall stands next to several spare segments. PHOTO: W. M. KECK OBSERVATORY AND ANDREW HARA

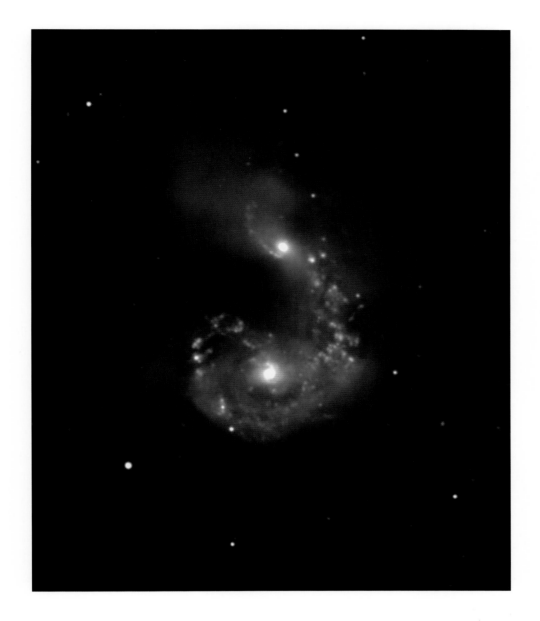

Henry David Thoreau once said, "I cannot make my days longer so I strive to make them better." The same is true, in a sense, for telescopes on Maunakea. Once built, a telescope's mirror size is fixed, but new instruments can dissect the collected light in ever more powerful ways.

MOSFIRE, or the Multi-Object Spectrograph for Infrared Exploration, is the newest instrument installed on the Keck I telescope. Developed by teams from UCLA, Caltech, UC Santa Cruz, and Keck Observatory, the $14 million, 5-ton instrument first opened its infrared-sensing eyes in 2012. This color composite image shows an interacting pair of galaxies, NGC 4038 and NGC 4039, better known as the Antennae Galaxies. The reddish blobs are large star-forming clusters, which are hidden from sight in normal visible light images. IMAGE: W. M. KECK OBSERVATORY AND IAN MCLEAN

New Instruments,
New Insights

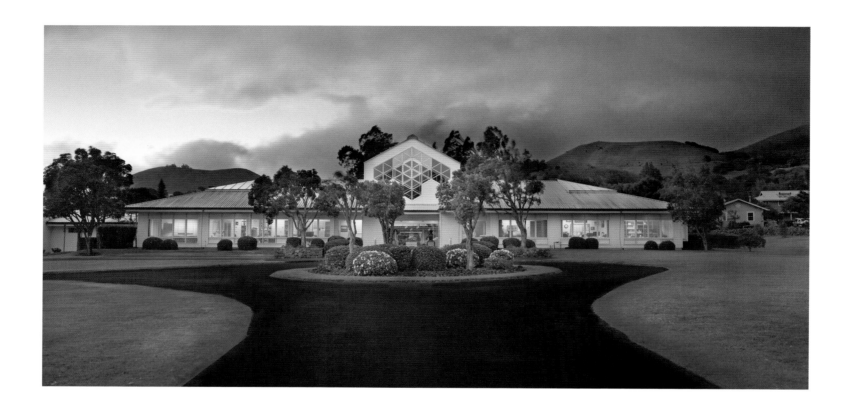

Although the demand for telescope time on Maunakea is greater than ever, fewer astronomers travel there today than in the past.

Thanks to technology, many telescopes and their instruments can now be controlled remotely. Astronomers using the Keck Observatory, for example, do so from the observatory's headquarters in the town of Waimea or from one of several control rooms at universities in California and as far away as Australia. Subaru Telescope offers astronomers the option of remote observing at the telescope on Maunakea or remotely from its Hilo headquarters or from Japan. And astronomers awarded time with NASA's Infrared Telescope Facility can control its instruments from anywhere in the world via the Internet.

Some observatories also offer "flexible scheduling" or "queue observing," in which an astronomer's desired observations are carried out by others. This type of observing provides great flexibility, allowing data to be collected when conditions are optimal rather than scheduling the observations for specific nights many months in advance. Implementations of this observing mode vary between observatories: at the James Clerk Maxwell Telescope (JCMT), for example, visiting astronomers carry out observations for other astronomers when the atmospheric conditions are not suitable for their own project, whereas queue observations at the Gemini North telescope and the Canada-France-Hawai'i Telescope (CFHT) are routinely executed by observatory staff.

The revolutionary design of the Keck I and Keck II telescope mirrors—each made from thirty-six individual segments—is reflected in the design of the observatory's headquarters in Waimea. From here, astronomers observe the nearest and farthest reaches of the universe without enduring the physical hardships of working in the thin air on Maunakea's summit at nearly 14,000 feet. PHOTO: W. M. KECK OBSERVATORY AND RICK PETERSON

Observing from Afar

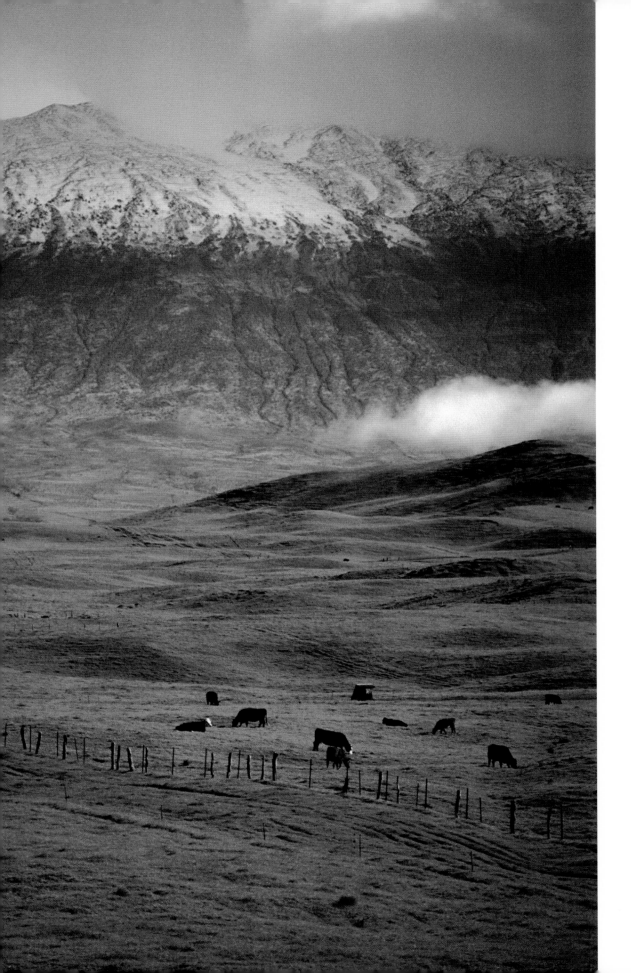

Peaceful Waimea

The picturesque town of Waimea on Hawai'i Island is home to the headquarters of Keck Observatory and the Canada-France-Hawai'i Telescope. PHOTO: BARBARA SCHAEFER

Two Generations
Side by Side

As twilight falls, stars begin to fleck the sky above the Gemini North telescope, seen in the foreground, and Canada-France-Hawai'i Telescope (CFHT). These two observatories, each of which has made pioneering contributions to astronomy, provide a vivid snapshot of how telescope design has evolved on Maunakea.

CFHT's 11.8-foot (3.6-meter) mirror was constructed from a thick slab of glass weighing 14 tons. The massive mirror's heft and rigidity help to maintain its correct shape. The entire telescope, which tips the scale at 325 tons, is housed in a dome 125 feet (38 meters) tall. When completed in 1979, it was the sixth largest telescope in the world.

Built two decades later, Gemini North represents a newer generation of telescopes that use thinner, more lightweight mirrors supported by computer-controlled hydraulic systems called active optics to maintain the mirror's ideal shape. Despite having five times more light-collecting area than CFHT's mirror, Gemini North's 26.2-foot (8.0-meter) mirror weighs only 50 percent more. The Subaru Telescope and the United Kingdom Infrared Telescope (UKIRT) use a similar thin mirror design. The Gemini North telescope, which weighs 342 tons, fits within a dome that stretches 150 feet (46 meters) above ground, comparable in size to CFHT's dome. PHOTO: MICHAEL WEST

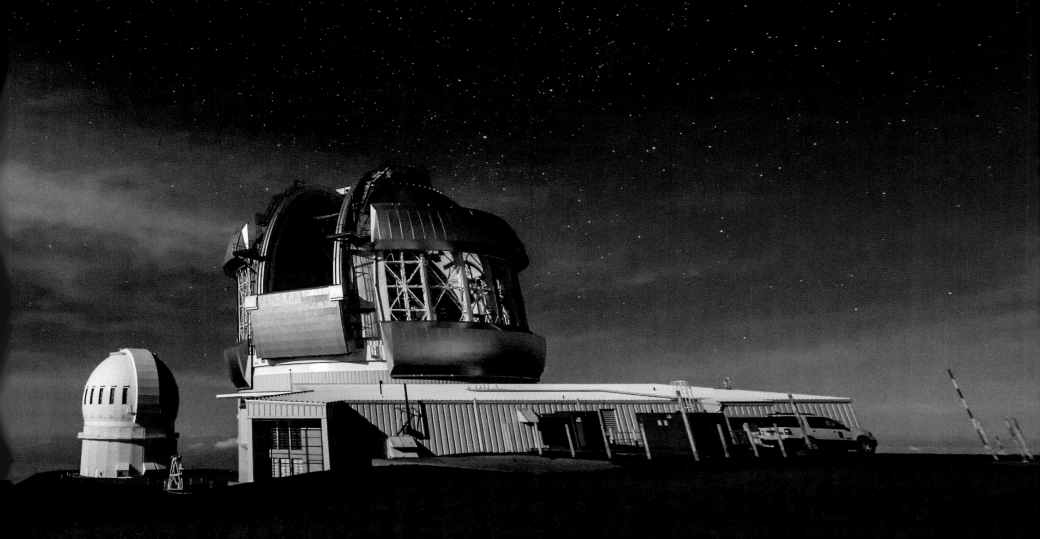

Inside Gemini North

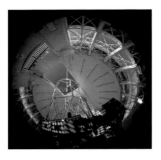

The interior of the dome that houses the Frederick C. Gillett Gemini telescope, more commonly known as the Gemini North telescope, is seen in this fisheye view. Gillett, who died in 2001, was a pioneer in infrared astronomy and a key figure in bringing Gemini Observatory from dream to reality.

The dome's vents visible in the photo allow air to flow across the telescope, equalizing temperatures inside and outside, which reduces localized turbulence that can blur images. The massive dome can do a complete revolution in just two minutes, allowing the telescope to be quickly aimed at different regions of the sky. The telescope, which weighs nearly 400 tons, rides on oil bearings and is balanced so precisely that it can be pushed by a single person when drive motors are turned off. PHOTO: GEMINI OBSERVATORY/AURA

Transparent Dome

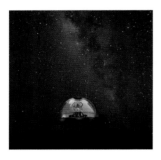

This unusual image shows the Gemini North telescope in 2004 with what appears to be a see-through dome. It's an optical illusion created by combining forty individual photographs of the open Gemini North dome while it rotated. The Milky Way stretches overhead. The telescope's dome, weighing nearly 400 tons, moves with the grace of a ballerina, turning almost silently as the telescope points at different regions of the sky. PHOTO: GEMINI OBSERVATORY, PETER MICHAUD, AND KIRK PU'UOHAU-PUMMILL

Mirror Support

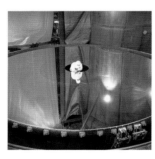

A major advance in telescope design over the past few decades has been the development of thinner, more lightweight mirrors. In the past, mirrors had to be made from thick slabs of glass whose stiffness helped preserve their shape. However, an innovative new design introduced in the 1990s allowed thinner, flexible mirrors to be used by telescopes, including the Subaru and Gemini Observatories on Maunakea.

The primary mirror of the Gemini North telescope, shown here, is a single piece of glass that's 26.2 feet (8.0 meters) in diameter yet only about 8 inches (20 centimeters) thick—about the same ratio of diameter to thickness as two CDs stacked on top of each other. Such a large but thin mirror isn't rigid enough to maintain its shape without support, so it rests on a bed of 120 computer-controlled hydraulic actuators that gently nudge the mirror into optimal shape in response to shifting gravitational pulls as the telescope tracks across the sky.

To focus light properly, a telescope's mirror must also be extremely smooth, otherwise irregularities on its surface won't reflect light striking them in the right direction. To focus visible and infrared light, the mirror must be smoother than a tiny fraction of the thickness of a human hair, an incredibly demanding engineering feat. The Subaru Telescope's 26.9-foot (8.2-meter) mirror, for example, has been polished so smoothly that irregularities on its surface are typically only 0.0000005 inches (0.000012 millimeters) or less. To put it in perspective, if the Subaru or Gemini mirrors were expanded to the size of the entire Big Island of Hawai'i, the largest peaks or valleys on the glassy landscape would only be about the thickness of a sheet of paper. PHOTO: GEMINI OBSERVATORY/AURA

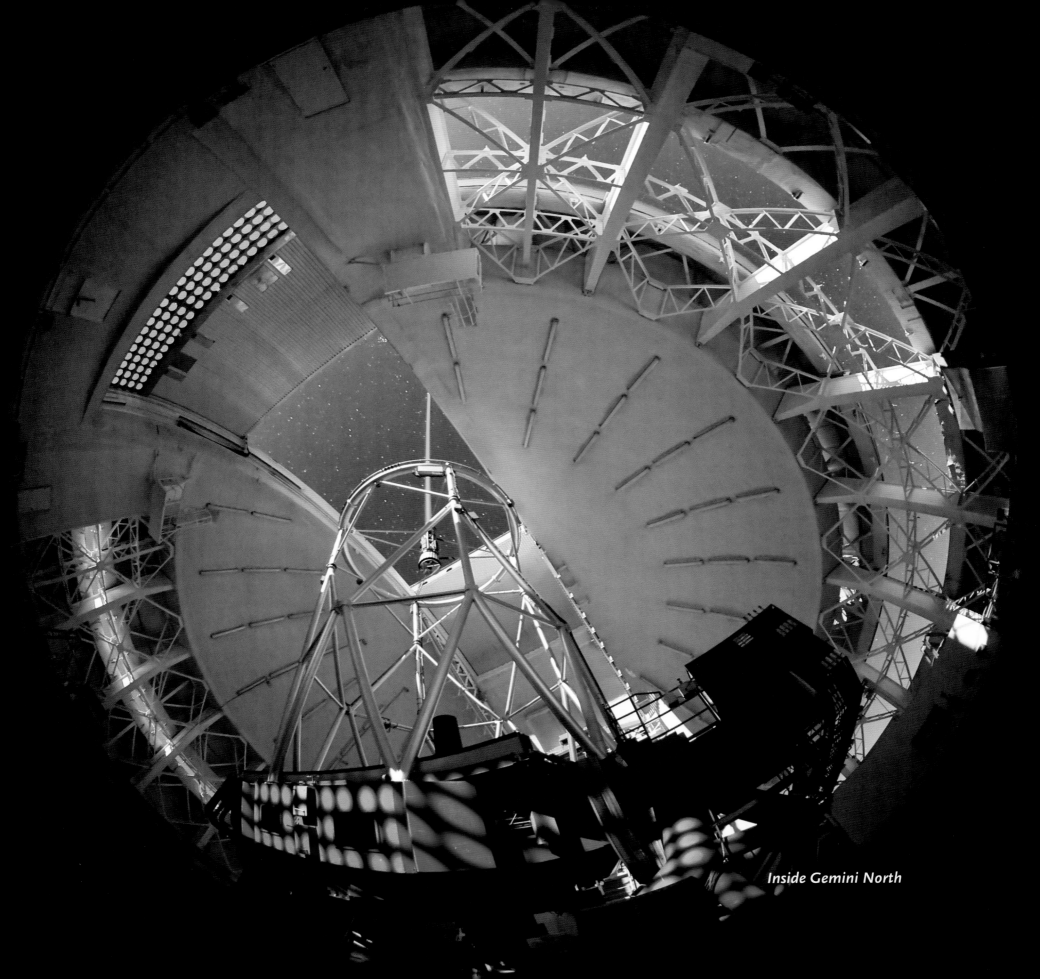

Inside Gemini North

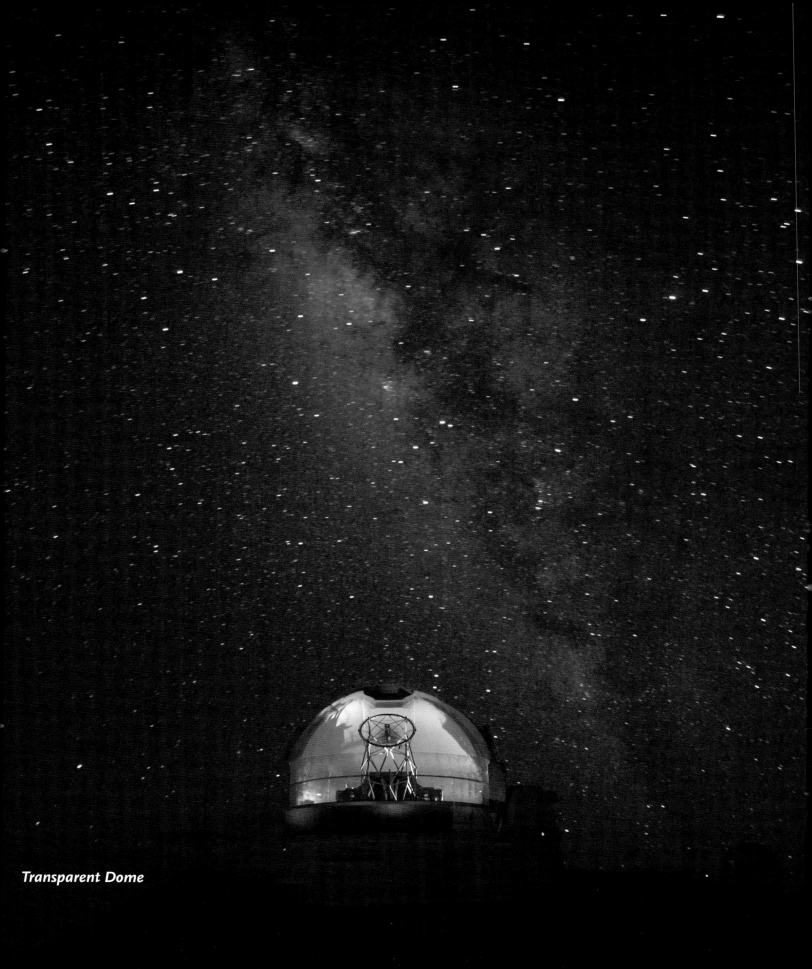

Transparent Dome

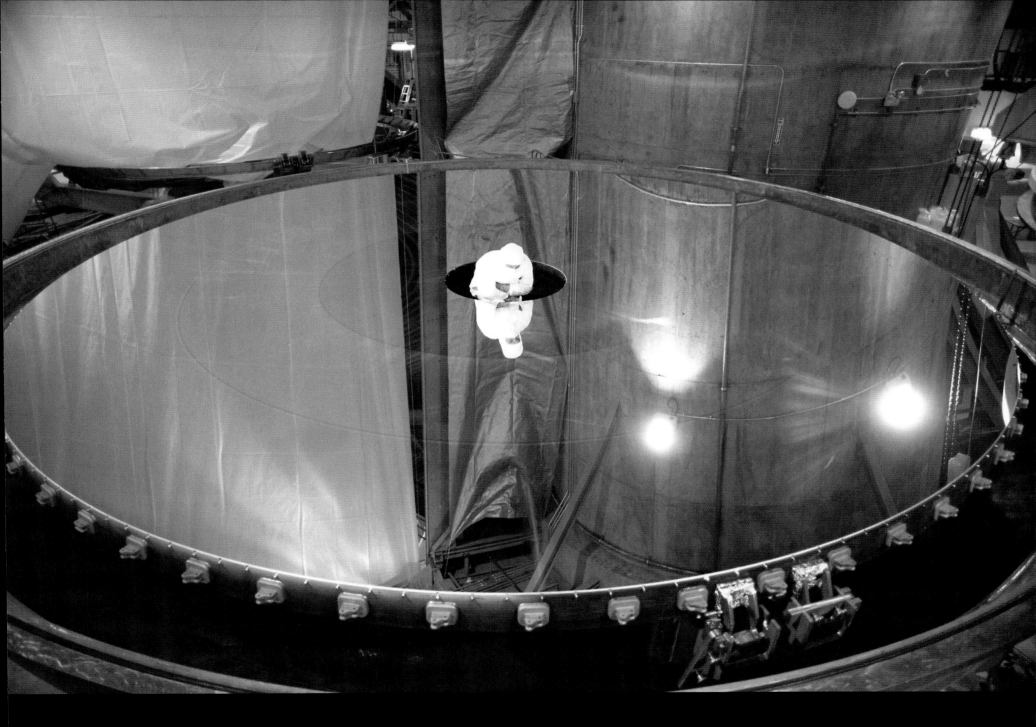

Mirror Support

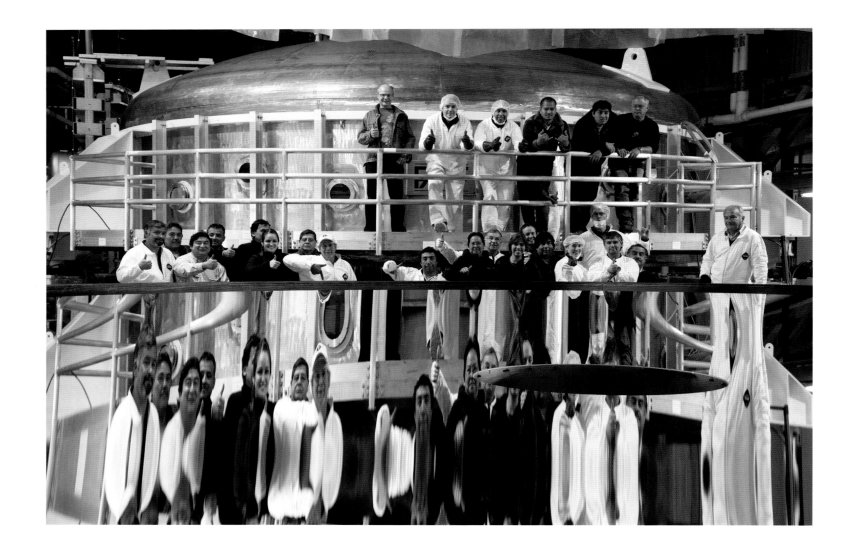

Reflecting on a Job Well Done

Despite frequent cleaning, a telescope's mirror requires complete recoating every few years as its reflective surfaces tarnish. The mirror—which weighs tons—must be gently removed from the telescope and placed inside a specially designed coating chamber. There, massive machinery and sophisticated electronics are used to deposit a microscopically thin layer of reflective material on the glass.

Most telescope mirrors are coated with aluminum, prized for its high reflectivity. Beginning in 2004, however, the Gemini Observatory switched to a silver coating because experiments showed that it reflects infrared light more efficiently than aluminum. Less than 2 ounces (about 50 grams) of silver are deposited onto the telescope's 26.2-foot (8.0-meter) mirror in a layer whose thickness is less than that of a human hair. Tests of the new silver coating confirmed that it reflects nearly 99 percent of infrared light that strikes it, while also reflecting visible light very well.

Here Gemini engineers and technicians pose next to the newly recoated Gemini North telescope's mirror. The coating chamber is visible in the background. PHOTO: GEMINI OBSERVATORY/AURA AND MANUEL PAREDES

The heart of every telescope is its light-collecting mirror, but its brain is the sophisticated suite of instruments that analyze the light to reveal its secrets. Here John White, senior instrumentation engineer at Gemini Observatory, works to install a new instrument on the Gemini North telescope known as the Gemini Near Infrared Spectrograph, or simply GNIRS. PHOTO: GEMINI OBSERVATORY/AURA

Distilling Light

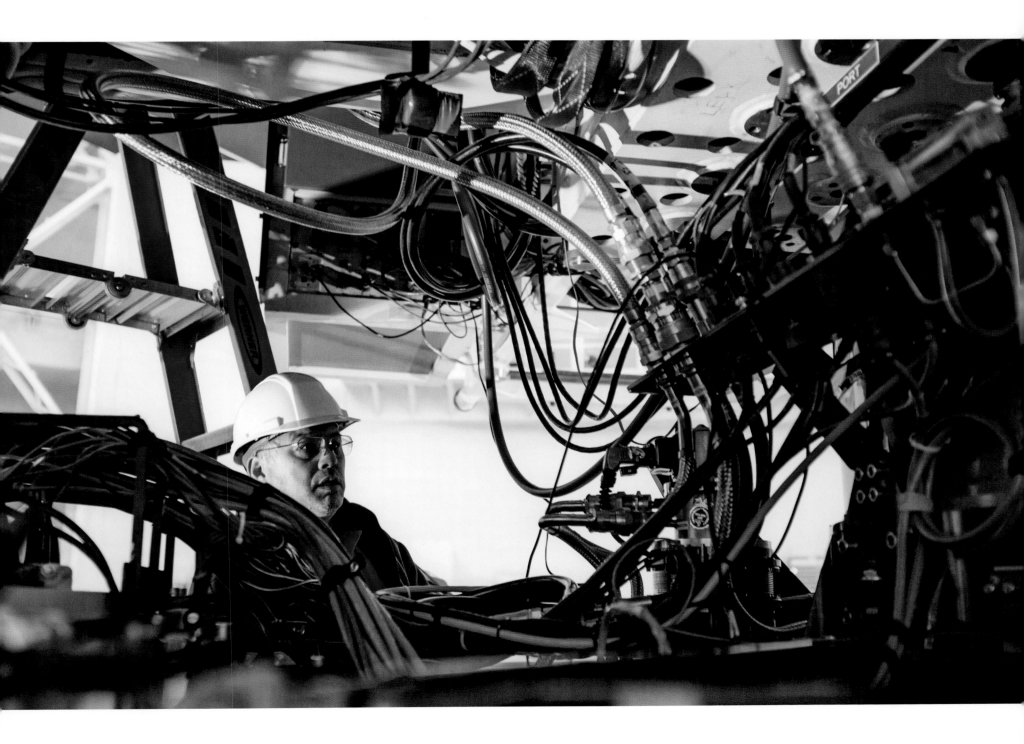

The Light of Three Realms

"The true mystery of the world is the visible, not the invisible," said the Irish writer Oscar Wilde.

This nighttime view towards the Gemini North telescope captures light from three different sources: the Earth, the cosmos, and people. To the left, lava flowing from Kīlauea volcano glows beneath the clouds that blanket the Big Island's lower elevations. Stars appear as luminous streaks in the clear skies overhead, their apparent motion caused by our planet's rotation during this long exposure. A trail of car lights is also visible, evidence of human activity on Maunakea during the night. PHOTO: RICHARD WAINSCOAT

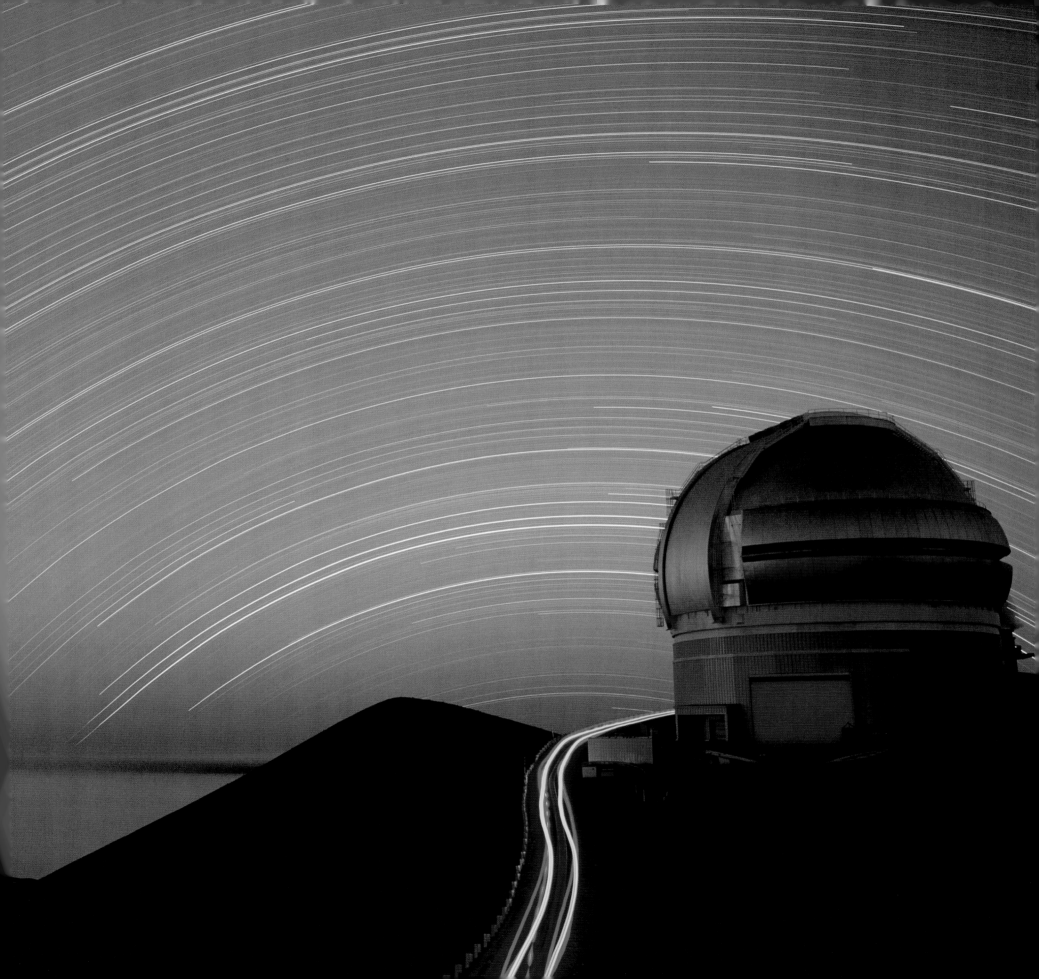

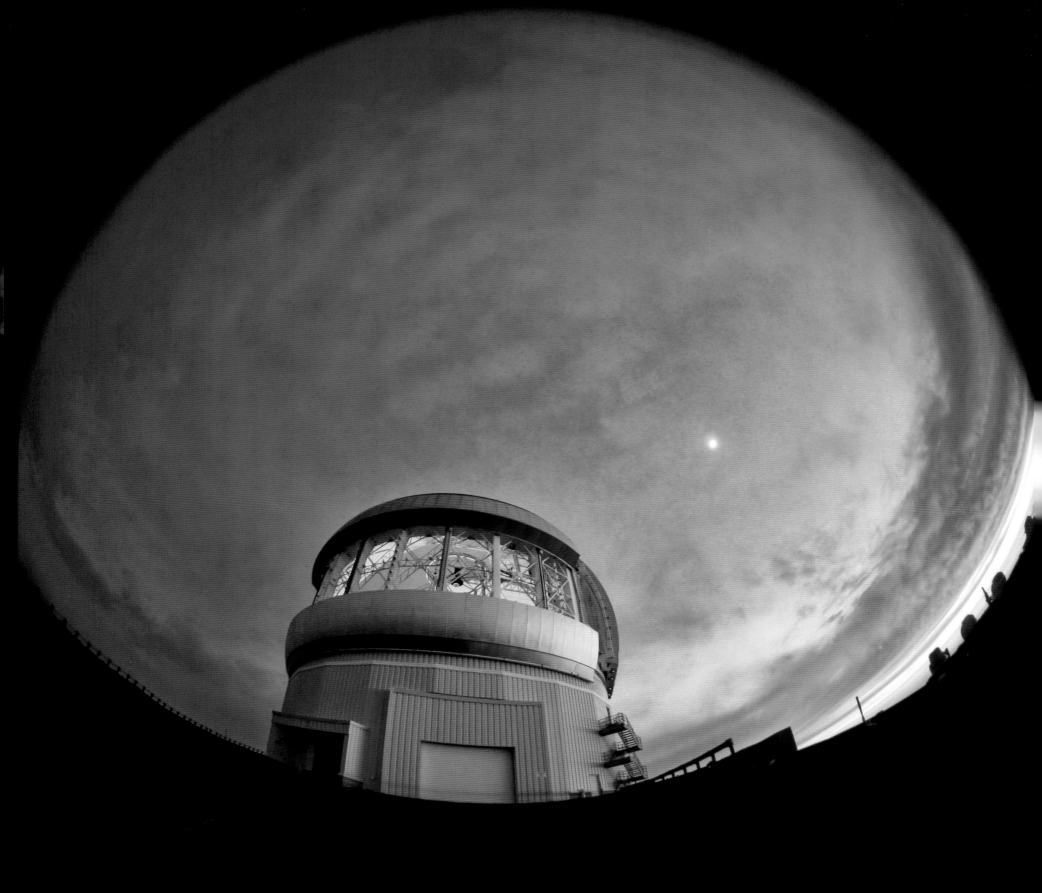

The Gemini North telescope waits for the sky over Maunakea to grow dark and another night of cosmic exploration to begin.

The cost and technological challenges of building cutting-edge telescopes like Gemini North have fostered scientific partnerships between nations. Gemini Observatory was conceived as a joint venture between the astronomical communities of the United States, the United Kingdom, Canada, Brazil, Argentina, Australia, and Chile, a partnership that has since evolved. Astronomers in each of Gemini Observatory's partner countries are allotted observing time in proportion to their nation's financial contribution to building and operating the telescopes, with astronomers at the University of Hawai'i and Chile also receiving time allocations as the host sites for Gemini North and Gemini South, respectively. PHOTO: GEMINI OBSERVATORY/AURA

Waiting for Nightfall

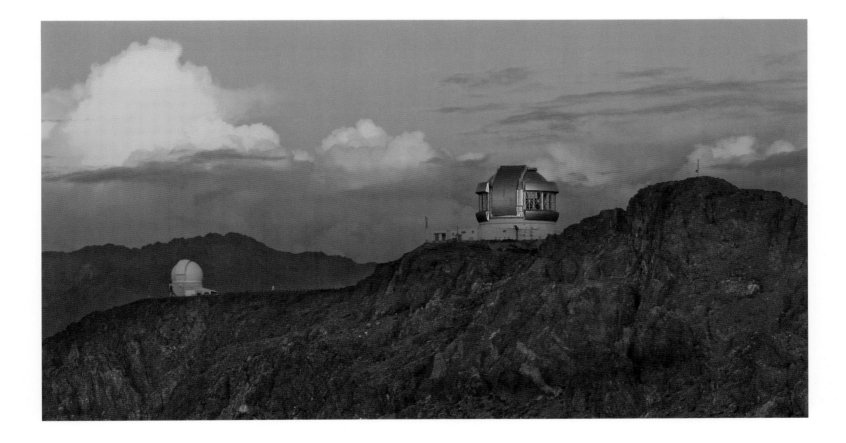

A Southern Sibling

Like the mythological characters for which it's named, the Gemini North telescope has a twin. Located in the Andes Mountains of Chile, the Gemini South telescope is the same size and design as its northern sibling on Maunakea. Each twin has a different suite of instruments, although some instruments have occasionally been swapped. Having telescopes in both the northern and southern hemispheres allows Gemini Observatory to observe objects almost anywhere in the sky. PHOTO: GEMINI OBSERVATORY/AURA

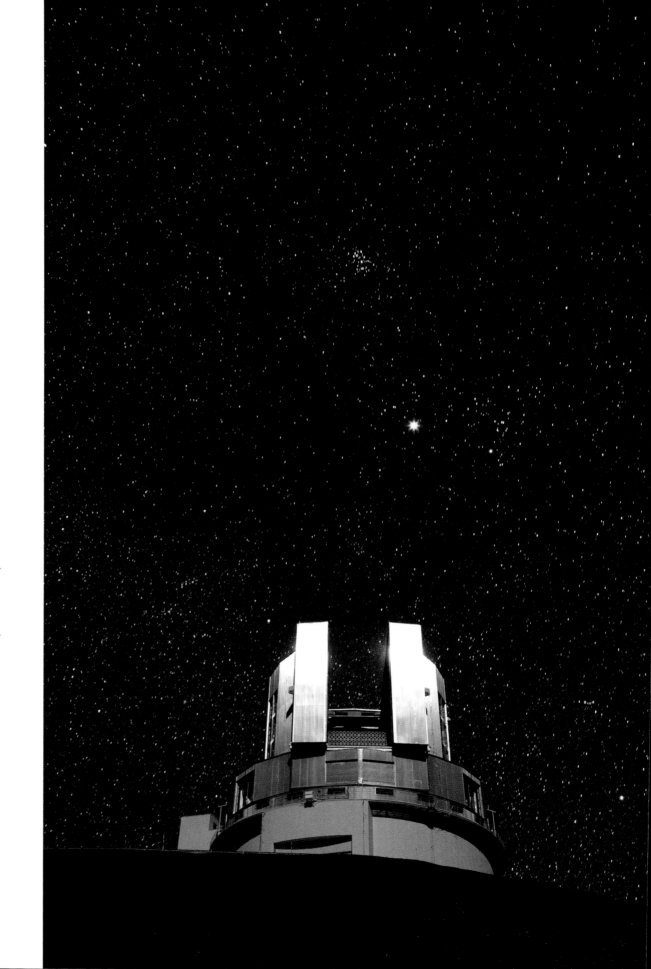

Subaru Telescope

Japan's most powerful optical and infrared telescope isn't found in the Land of the Rising Sun. It's 4,000 miles away on the summit of Maunakea. Named for a prominent cluster of stars known as the Pleiades in English or Makaliʻi in Hawaiian, the Subaru Telescope has been capturing breathtaking views of the heavens since 1999. The heart of the telescope is an ultrathin glass mirror 26.9 feet (8.2 meters) in diameter but only 8 inches (20 centimeters) thick.

Here the Subaru Telescope is seen against a starry background that includes its namesake Pleiades star cluster. The National Astronomical Observatory of Japan owns and operates the telescope. PHOTO: SUBARU TELESCOPE/NAOJ

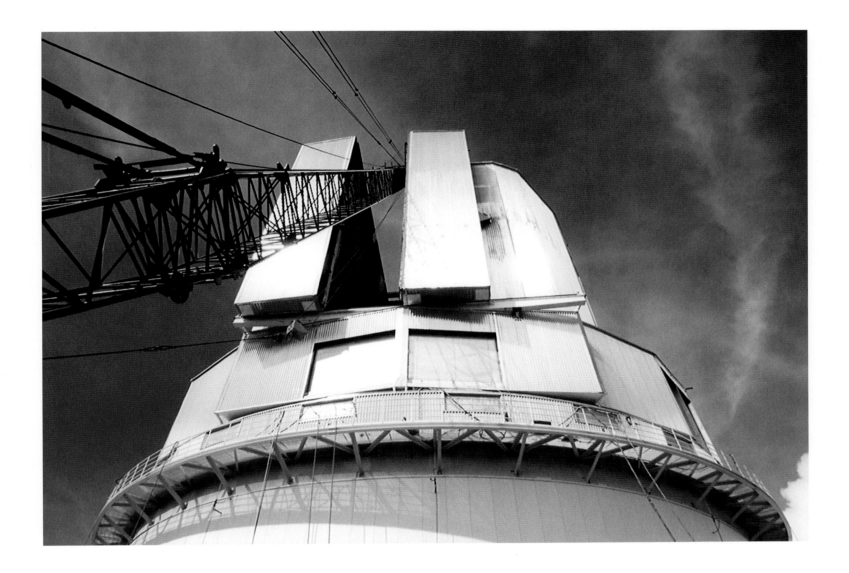

A Japanese proverb says 愚公山を移す, which means, "Nothing is impossible for those who have faith."

Construction of the Subaru Telescope on Maunakea began in 1992 and was completed in 1998. In April 1997, the National Astronomical Observatory of Japan established a base facility in the Big Island city of Hilo to support construction and operation of the telescope.

Subaru's builders faced many challenges. Heavy equipment and parts had to be transported up and down the mountain without interfering with the operation of other observatories. Workers had to contend with the effects of oxygen deprivation and dehydration in the rarefied air above the clouds. Control systems for the telescope and dome had to work with incredible precision.

This photo shows the Subaru Telescope being carefully installed inside its protective dome. The telescope's first scientific observations in January 1999 were the culmination of years of hard work, a testament to its builders' tenacity and to the human spirit of exploration that makes us reach for the stars. PHOTO: SUBARU TELESCOPE/NAOJ

Reaching for the Stars

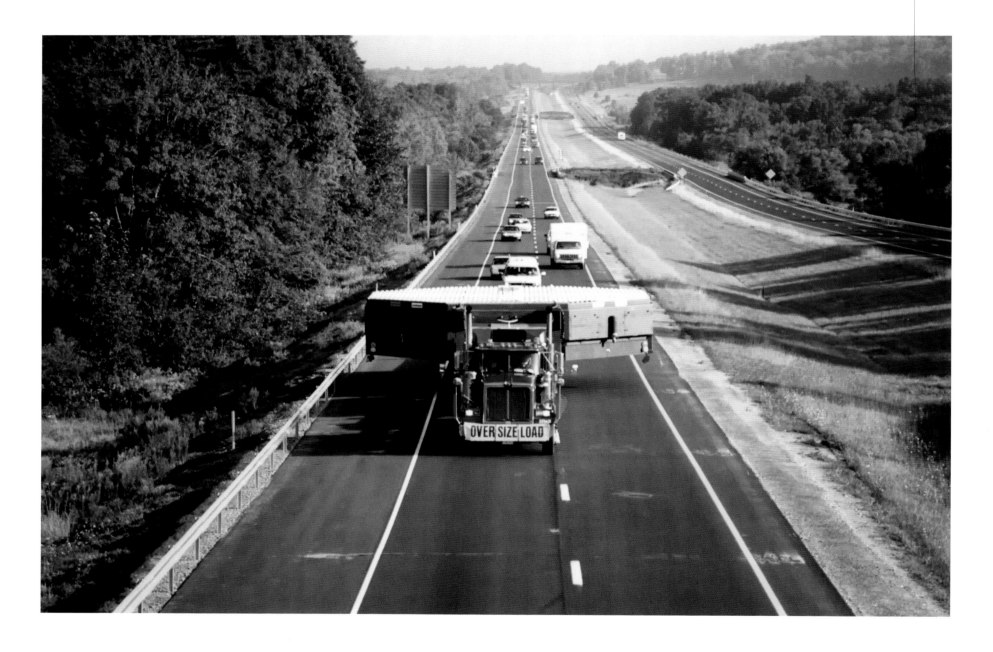

The Economy of Astronomy

Astronomy has become a major industry in Hawai'i, an economic engine that currently pumps more than $150 million dollars annually into the local economy on the Big Island and employs about six hundred people in astronomy-related jobs. More than $1 billion dollars have been invested in the infrastructure of the thirteen Maunakea observatories, and that amount will more than double with the construction of the Thirty Meter Telescope.

Here the Subaru Telescope's 26.9-foot (8.2-meter) mirror is seen on its 1994 journey from New York where it was cast to Pennsylvania where it was polished, before heading on to Maunakea. PHOTO: SUBARU TELESCOPE/NAOJ

Like the charming sea turtles that swim along Hawai'i's shores, Maunakea's telescopes have shells for protection.

Freezing temperatures, hurricane-force winds, and occasional blizzards aren't conditions usually associated with Hawai'i, but they're found on Maunakea—the only place in the state to have a snowplow. Fierce and sudden storms can engulf the mountain during winter, requiring quick evacuation. Wind speeds of more than 125 miles per hour have been recorded at the summit.

Shielded from the elements, the telescope peers through a narrow slit to view the night sky. Because temperature differences inside and outside the dome can create pockets of air turbulence that distort images of planets, stars, and galaxies, domes are well ventilated to allow air temperatures to stabilize. A coat of white paint or other reflective material helps prevent the fierce tropical sunlight from being absorbed and heating up the interior of the dome during the day. The interiors of many domes are also cooled to the anticipated nighttime temperature to further reduce temperature differences inside and outside the dome.

The Subaru Telescope's dome, shown above, has a cylindrical shape rather than the more traditional round one. This shape reduces turbulence near the telescope by helping air flow more easily around the dome, yielding better image quality. PHOTO: SUBARU TELESCOPE/NAOJ

A Protective Shell

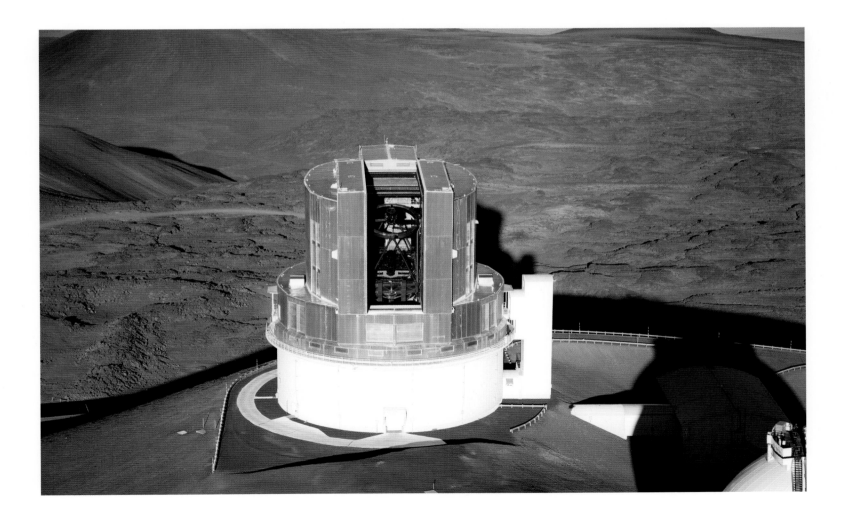

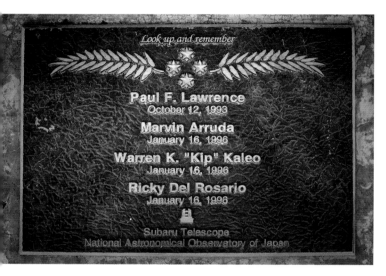

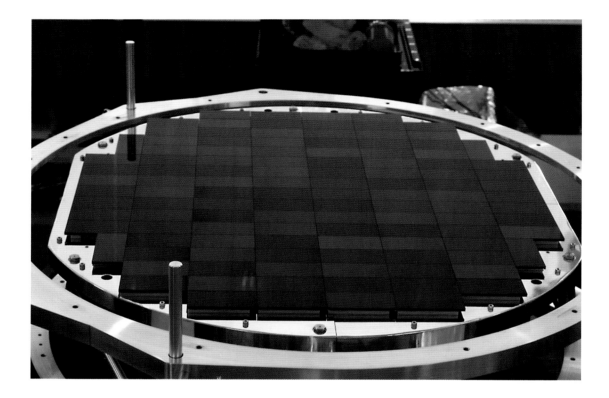

In Memory

Tragedy struck not once but twice during construction of the Subaru Telescope, taking the lives of four workers. This plaque, located outside the Subaru Telescope on Maunakea, honors their sacrifice. PHOTO: SAEKO HAYASHI AND SUBARU TELESCOPE/NAOJ

Paparazzi of the Stars

Photography as we know it began in 1822 when a Frenchman named Joseph Nicéphore Niépce recorded a permanent image by exposing a pewter plate coated with bitumen and lavender oil to light. He called the process "heliography," or "sun writing."

Astronomers quickly realized the power of photography for their own work. In 1840, John Draper took the first successful photograph of the moon. Two years later, Gian Alessandro Majocchi recorded the first image of a solar eclipse. The first photograph of a star—Vega—was taken at Harvard College's Observatory in 1850. Astronomy had entered a new era.

One of the most important technological developments in astronomy over the past half century has been the transition from photographic plates to more sensitive digital detectors known as CCDs (charge-coupled devices) for recording astronomical images. Whereas a photographic plate records only a small percentage of the light striking it, modern digital detectors capture nearly all the light.

Since their introduction in the 1980s, digital cameras used by Maunakea's telescopes have grown in size and sensitivity. The Canada-France-Hawai'i (CFHT) and Subaru telescopes in particular have been at the forefront of wide-field imaging. For years, CFHT's MegaCam was the biggest camera in the world, with 340 million pixels. Subaru Telescope's new Hyper Suprime-Cam (HSC), shown here, boasts a whopping 870 million pixels obtained by joining 116 individual CCDs together to create a gigantic camera that can image large areas of sky with exquisite resolution. Hyper Suprime-Cam saw first light in 2012. PHOTO: HSC PROJECT AND SUBARU TELESCOPE/NAOJ

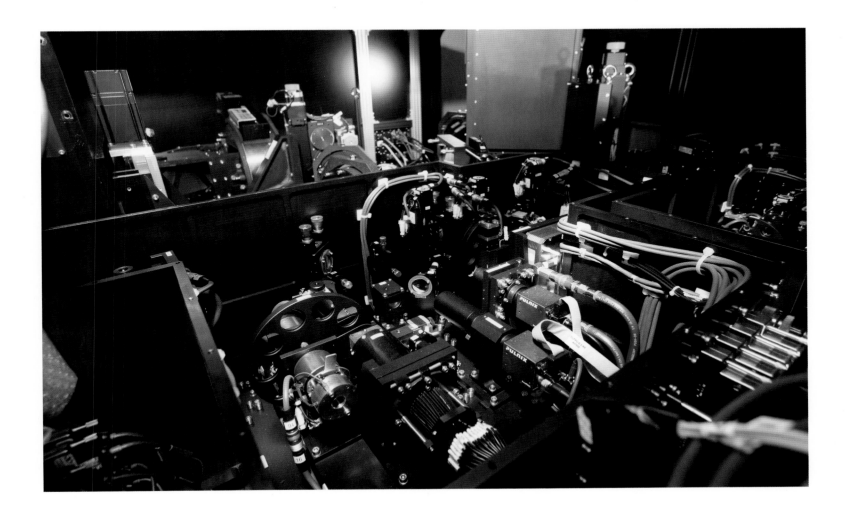

Astronomy has come a long way since Galileo Galilei's first telescopic observations in 1609. Back then, astronomers had to be artists as well as scientists, drawing whatever they saw in the sky. Telescopes have grown much bigger and more complicated since then, and sensitive instruments that record and analyze light in ways that Galileo could never have imagined have replaced the human eye.

The Subaru Telescope's adaptive optics system is a good example. This sophisticated system reduces image distortions that occur whenever light passes through our planet's turbulent atmosphere. By continuously monitoring the light of a bright star or a laser-generated artificial one, the adaptive optics system redirects light to a small, deformable mirror whose shape is changed many times per second to precisely cancel out atmospheric distortions, yielding crisp images of planets, stars, galaxies, and other celestial objects.

The first adaptive optics system on Maunakea was a thirteen-element system developed by the University of Hawai'i's Institute for Astronomy in 1994. Continuous technological advancements have led to Subaru's current system with 188 corrective elements, which yield images as sharp or sharper than those obtained with the Hubble Space Telescope orbiting far above Earth's troublesome atmosphere. Other optical and infrared telescopes on Maunakea use similar adaptive optics systems. PHOTO: HIDEAKI FUJIWARA AND SUBARU TELESCOPE/NAOJ

A Marvel of
Engineering

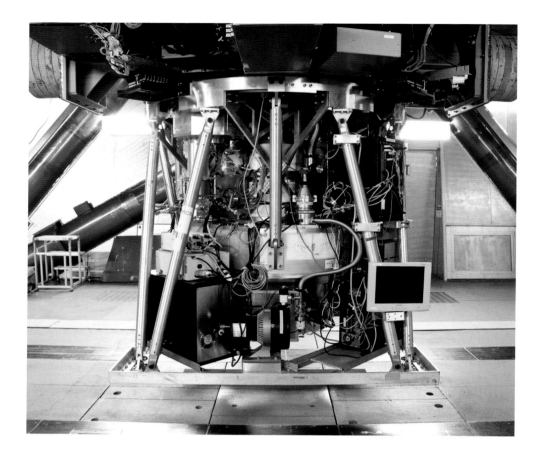

Most of our knowledge of the universe comes from light. Each telescope on Maunakea has its own unique arsenal of sophisticated instruments that extract information from the light collected by the telescope's mirror. These instruments can weigh as much as a car, yet they must operate with microscopic precision. An example is the Subaru Telescope's Multi-Object InfraRed Camera and Spectrograph (MOIRCS), seen here attached to the bottom of the telescope. PHOTO: SUBARU TELESCOPE/NAOJ

Learning from Light

A Sky Filled with Stars

On a moonless night, far from city lights, a person with good eyesight can see approximately 2,000 to 2,500 stars at any moment. Every one of them is part of our Milky Way galaxy, and all are more luminous than the sun. Without telescopes, we can see only the nearest stars, a minuscule portion of our cosmic home. The most distant star visible to the unaided eye is V762 Cas, located about 16,000 light-years from Earth. But telescopes reveal much more distant stars; the most remote stars in the Milky Way are almost 100,000 light-years away. PHOTO: MICHAEL WEST

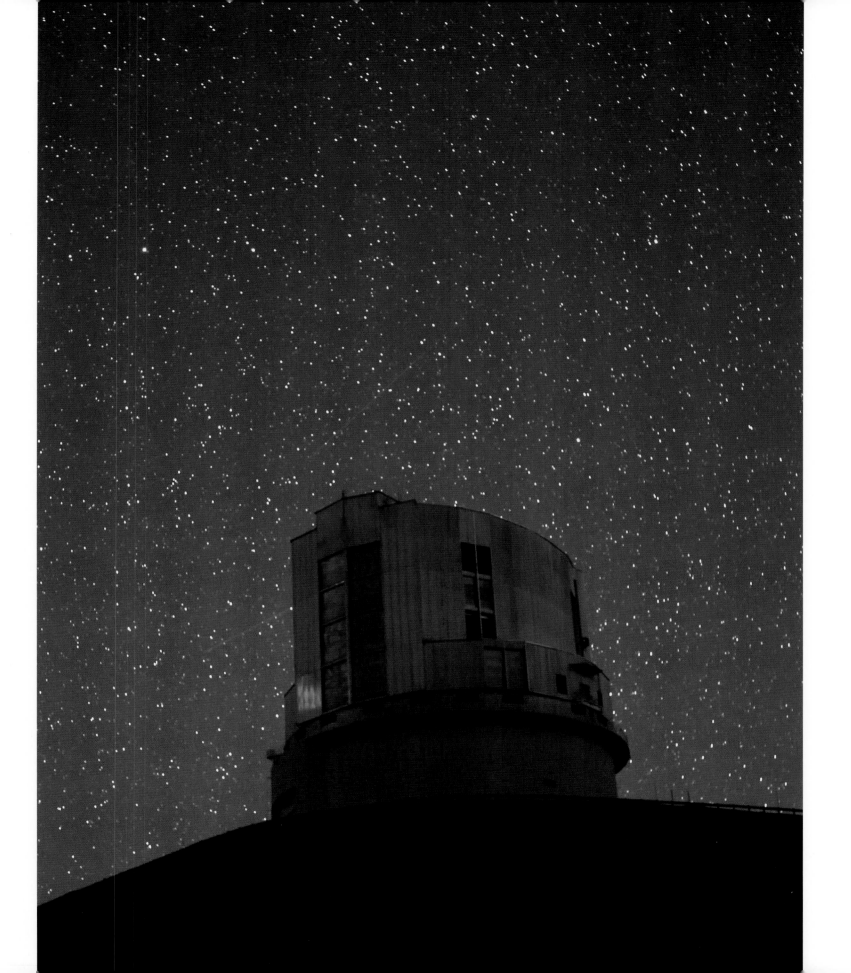

A Subaru Sunset

The popular image of an astronomer peering through a telescope eyepiece is a thing of the past. Today, professional astronomers almost never look through telescopes because the human eye detects only a tiny fraction of the light it receives. This problem is exacerbated by Maunakea's nearly 14,000-foot elevation, where the reduced oxygen dramatically affects night vision. Astronomical observations are now made with cameras equipped with electronic detectors that are much more sensitive than the human eye and allow data to be stored, displayed, and analyzed on powerful computers.

A rare exception occurred at the inauguration of the Subaru Telescope in 1999, when an eyepiece was temporarily installed to allow Princess Norinomiya Sayako of Japan to look through the telescope directly, truly a once-in-a-lifetime experience.

PHOTO: MICHAEL WEST

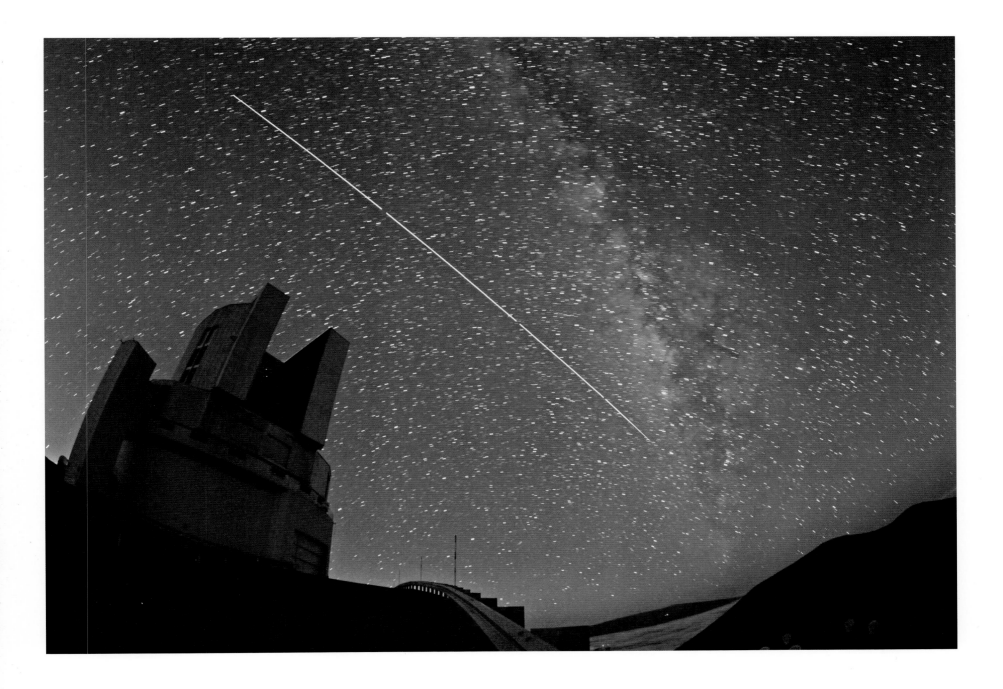

The spirit of human exploration is beautifully captured in this May 2014 photo of the Subaru Telescope as the International Space Station (ISS) streaks overhead between it and the Milky Way.

In command of the ISS was Japanese astronaut Koichi Wakata. The famed astronaut had visited Hilo three years earlier, giving a public lecture to a packed house at the ʻImiloa Astronomy Center of Hawaiʻi and returning a commemorative banner to Subaru headquarters that he had carried into space during a 2009 mission to the space station. PHOTO: HIDEAKI FUJIWARA AND SUBARU TELESCOPE/NAOJ

Between Heaven
and Earth

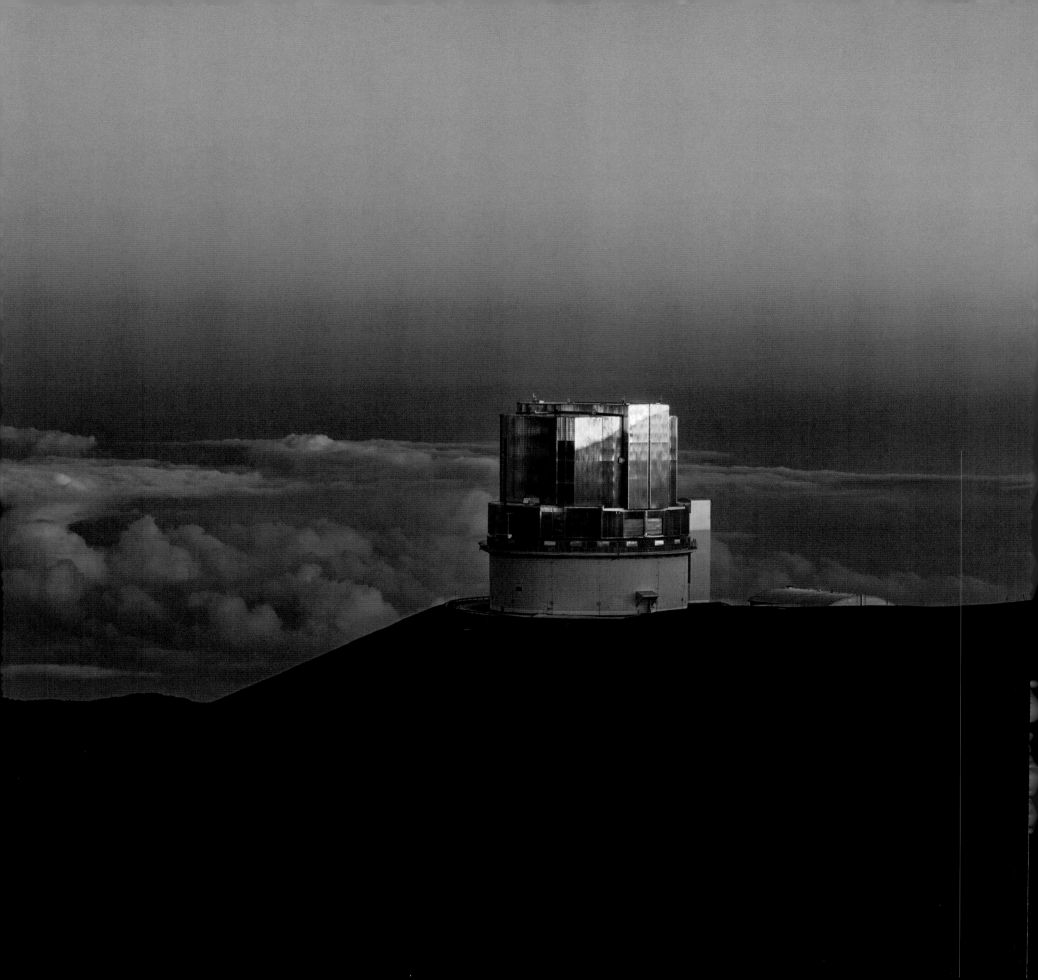

The First Rays of Sunrise

The first rays of sunlight glint off the dome of the Subaru Telescope as a new day dawns. PHOTO: MICHAEL WEST

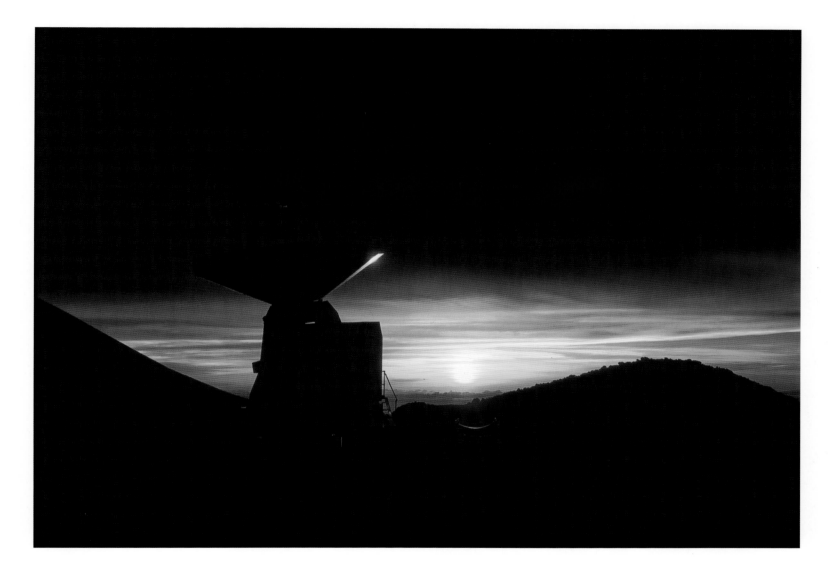

The Submillimeter Array (SMA) is a single telescope composed of eight separate receivers that work together to collect microwaves (millimeter and submillimeter light), the kind that comes from cold, dusty regions of the universe where planets, stars, and distant galaxies are born.

The SMA was built as a partnership between the Smithsonian Astrophysical Observatory in the United States, the Academia Sinica Institute of Astronomy and Astrophysics in Taiwan, and the University of Hawai'i's Institute for Astronomy, one of many successful international partnerships on Maunakea. Construction of the SMA began in 1995 and was completed in 2003. Each partner contributes to the maintenance and operation of the telescope. PHOTO: MICHAEL WEST

The Submillimeter Array

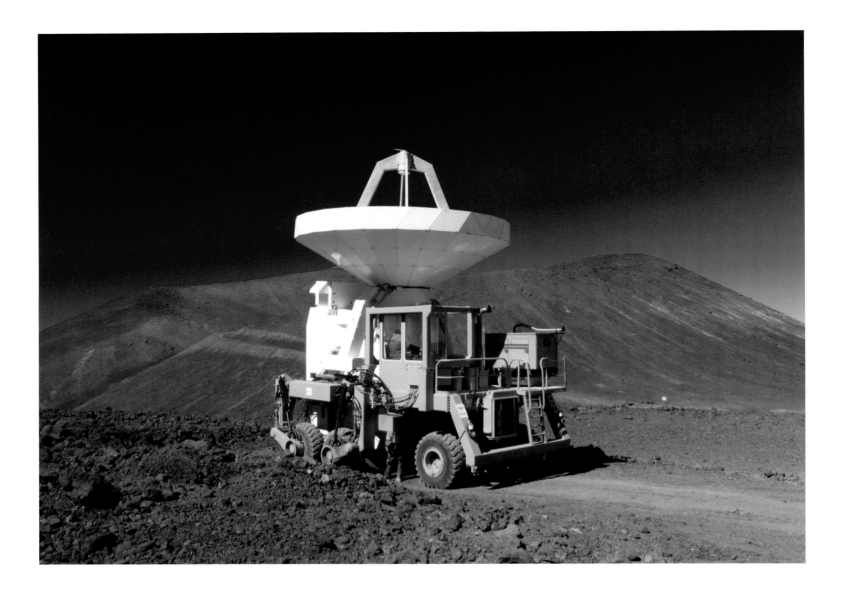

A Moveable Telescope

The eight dishes that make up the Submillimeter Array (SMA) work together like segments of a single giant mirror, but with an important difference—they're moveable. The dishes can be positioned at any of twenty-four different locations over an area of more than 50 acres (20 hectares)

The SMA produces images of the sky by combining light from all eight dishes, a process called interferometry. Changing the positions of the dishes allows the array to act like an enormous zoom lens. The SMA's vision is sharpest when the dishes are placed at their widest separation, providing very detailed views of small regions of the sky. Moving the receivers closer together, on the other hand, is like using a wide-angle lens that sees a larger region of the sky but with fewer small-scale details.

This photo shows one of the SMA's transporters preparing to move a dish on Maunakea. The specially designed transporters are quite agile on the rocky terrain, capable of repositioning several dishes per day. PHOTO: GLEN PETITPAS

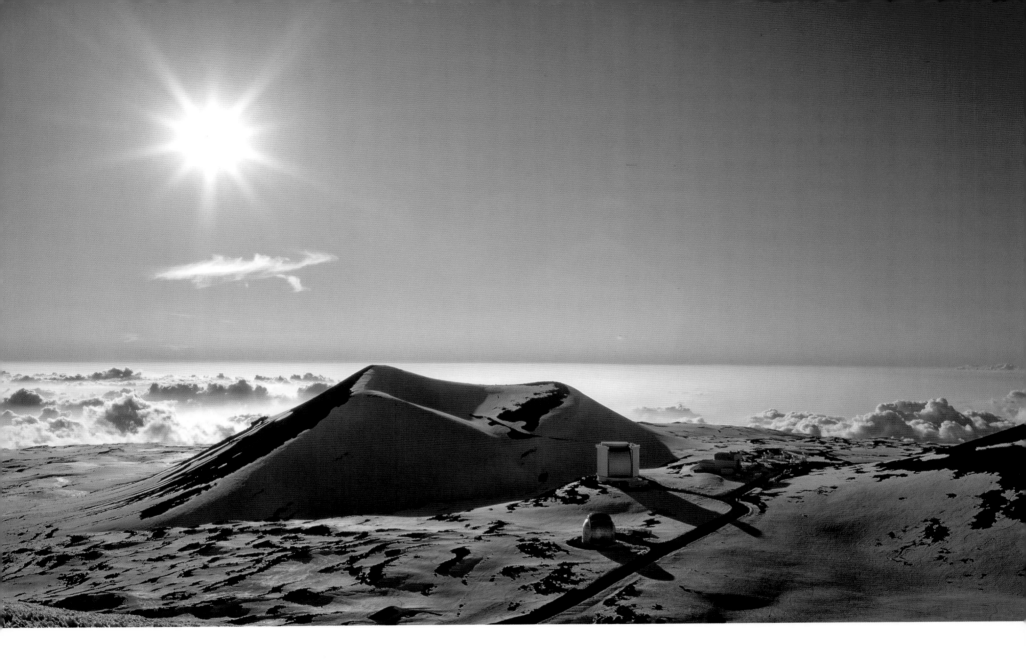

Infrared or submillimeter light that has traveled for millions or billions of years from a distant star or galaxy can be snuffed out in the final moments of its journey, absorbed by water vapor or carbon dioxide in our planet's atmosphere before it ever reaches telescopes on the ground. Maunakea's high altitude and dry air—its peak lies above 97 percent of the obscuring water vapor in our planet's atmosphere—make it an ideal site for infrared and submillimeter observations, allowing telescopes perched there to capture those photons before they're swallowed by the thicker moist layers of air below.

"Millimeter Valley" is the nickname given to the area on Maunakea that's home to the Caltech Submillimeter Observatory (CSO), James Clerk Maxwell Telescope (JCMT), and the Submillimeter Array (SMA). In the background is Pu'u Poli'ahu, named after the icy goddess of Maunakea, where a temporary telescope was placed in the 1960s to test the mountain's suitability for astronomical observations. PHOTO: MICHAEL WEST

Millimeter Valley

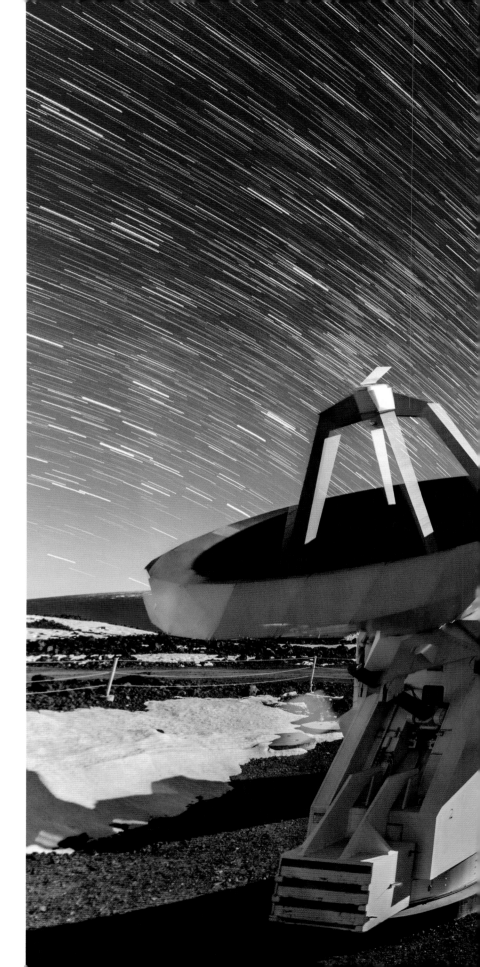

Playing with the
Light of the Universe

In his memoirs, the great Chilean poet Pablo Neruda recalled his visit to an astronomical observatory, where he "saw the writing of the stars for the first time" and watched in fascination as the telescope's instruments recorded "the palpitation of the stars in space, like an electrocardiogram of the sky."

Neruda would have loved Maunakea. This long-exposure photograph shows the Submillimeter Array (SMA) illuminated by stars and moonlight. The telescope consists of eight separate and moveable dishes, each 20 feet (6 meters) in diameter and weighing 50 tons (45 metric tonnes). The SMA forms high-resolution images of the sky by using a device known as a correlator to combine the light from each dish, revealing the palpitations of stars and galaxies across the universe. PHOTO: JASON CHU

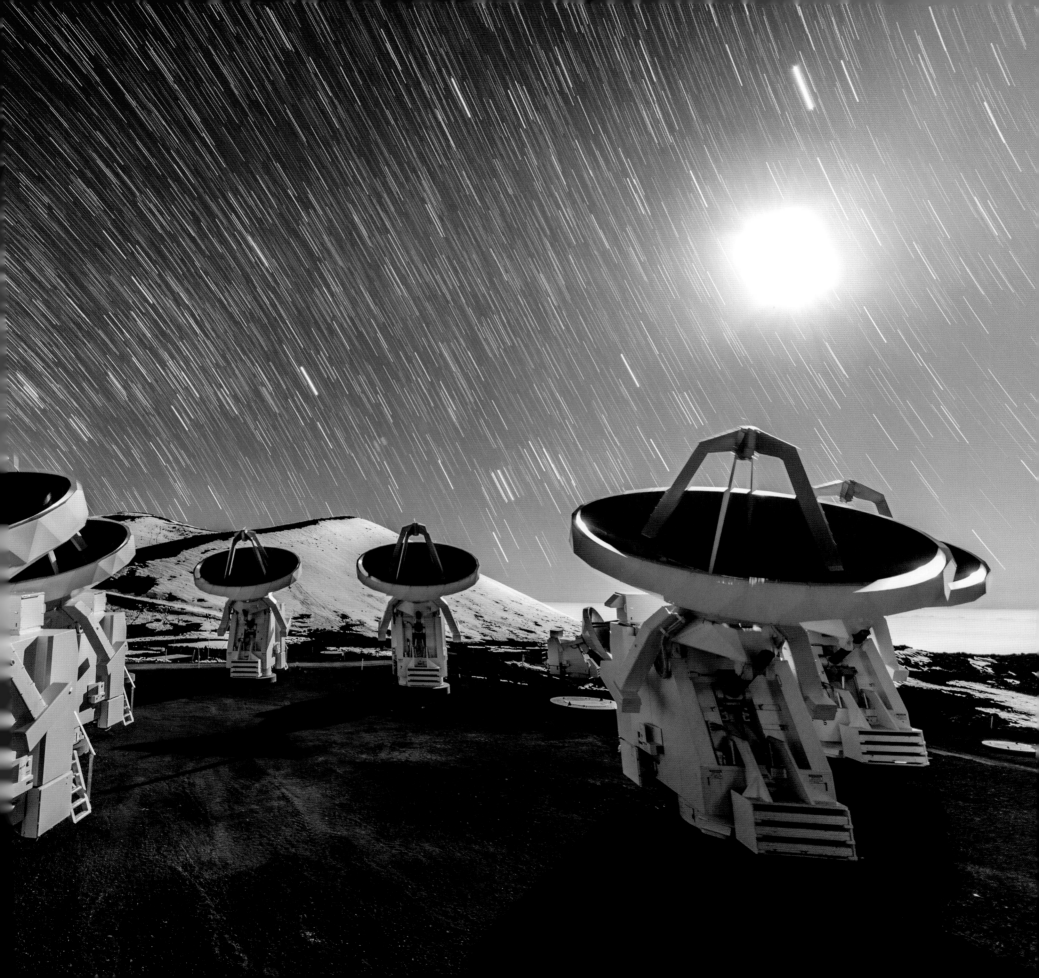

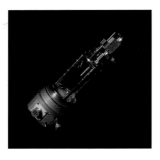

A Different Sort of Camera

Submillimeter telescopes—such as the Caltech Submillimeter Observatory (CSO), James Clerk Maxwell Telescope (JCMT), and Submillimeter Array (SMA) on Maunakea—collect light of much lower energies than the visible light our eyes can see. For this reason, they're designed differently from optical and infrared telescopes.

One of the biggest differences is instrumentation. Some, like SCUBA-2 on the JCMT, use devices called bolometers to detect light. A bolometer is, in essence, a thermometer that measures tiny temperature changes that occur when light strikes the telescope. Arrays of bolometers can be arranged like pixels in a digital camera to produce images or record spectra.

Other submillimeter instruments borrow techniques common to radio telescopes by using light-sensitive devices known as heterodyne receivers. These are similar to AM and FM radios, amplifying the electric field of incoming radio waves. One of the SMA's receiver inserts is seen here.
PHOTO: GLEN PETITPAS

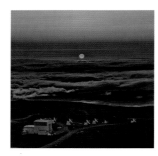

Superlative Science

During its first decade of operation, more than five hundred papers were published in leading scientific journals based on observations made with the Submillimeter Array (SMA). Collectively, the thirteen telescopes on Maunakea are among the most scientifically productive in the world.
PHOTO: MICHAEL WEST

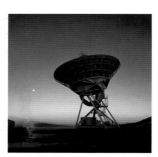

The Very Long Baseline Array

Just below Maunakea's summit, at an elevation of 12,000 feet (3,660 meters), sits a giant radio telescope. Built in 1992, it measures 82 feet (25 meters) in diameter and weighs 240 tons—as much as eight adult humpback whales. It's one of ten similar telescopes spread from the Virgin Islands across the United States to Hawai'i that make up the Very Long Baseline Array (VLBA). All ten VLBA dishes are controlled remotely from the National Radio Astronomy Observatory's Science Operations Center in New Mexico.

Radio telescopes are larger than other types for two reasons. First, radio waves have less energy than other types of light and hence larger dishes are needed to collect enough faint radio emission for detection. Second, radio images are intrinsically less detailed than those produced by other types of light, and this can be countered by building larger telescopes because they produce more detailed images.

Although the sun's intense glare prevents the visible light of planets, stars, and galaxies from being seen during the daytime, other types of light such as radio waves can be observed because the sun emits very little radio light. Here the VLBA observes radio emission from a distant galaxy during the early evening, while most other telescopes on Maunakea are still waiting for the sky to grow dark enough to reveal the stars. PHOTO: MICHAEL WEST

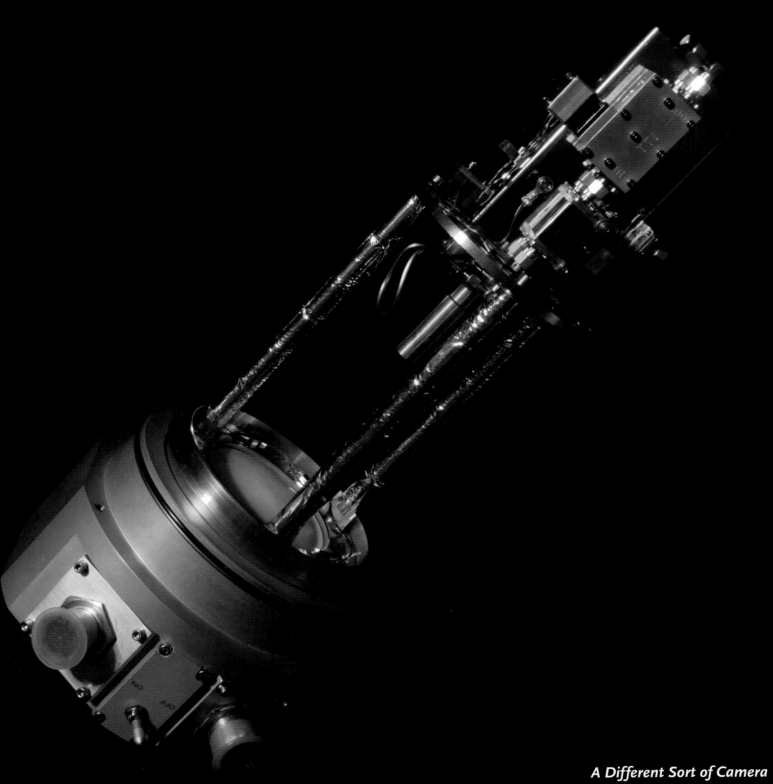

A Different Sort of Camera

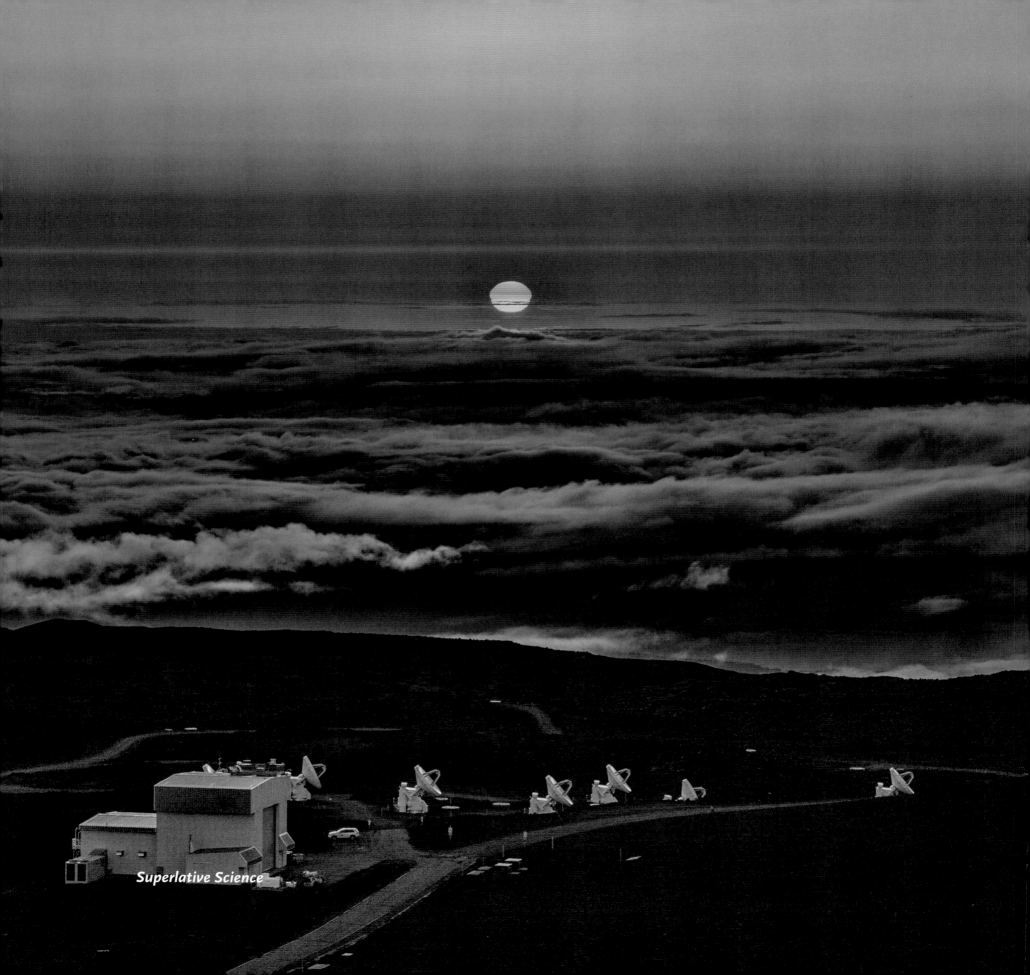

Superlative Science

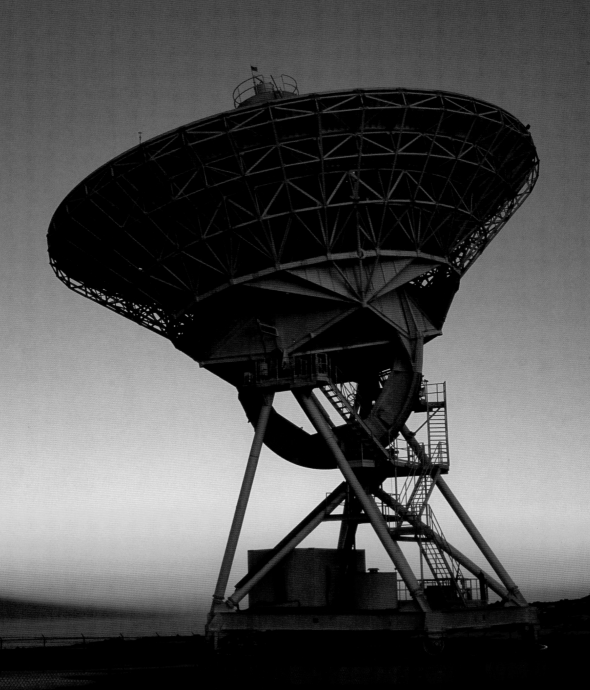

The Very Long Baseline Array

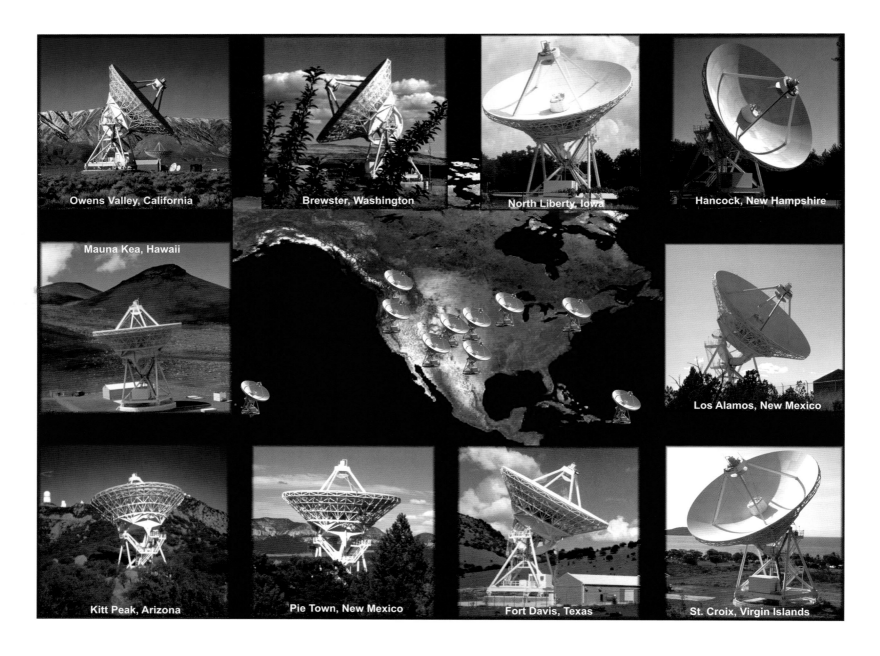

Owens Valley, California

Brewster, Washington

North Liberty, Iowa

Hancock, New Hampshire

Mauna Kea, Hawaii

Los Alamos, New Mexico

Kitt Peak, Arizona

Pie Town, New Mexico

Fort Davis, Texas

St. Croix, Virgin Islands

A Continent-Sized Telescope

The Very Long Baseline Array (VLBA) is, quite literally, a continent-sized telescope. Ten different radio dishes from St. Croix in the U.S. Virgin Islands to Maunakea in Hawai'i are operated from a center in Socorro, New Mexico, as part of the U.S. National Radio Astronomy Observatory.

By using the different VLBA dishes to observe the same object at the same time, it's possible to combine their data into a single image. In this way the VLBA effectively acts like a single radio telescope stretching over 5,000 miles (8,000 kilometers) from end to end. This allows it to see much finer details than any other telescope, including the Hubble Space Telescope. A person with the VLBA's sharp vision would be able to read a newspaper in Los Angeles while standing in New York. PHOTOS: NRAO/AUI/NSF

Human beings have always watched the skies, so it's not surprising that astronomy is the oldest of all sciences, with roots that go back to the dawn of civilization. Two millennia ago, the great Greek philosopher Plato said, "Astronomy compels the soul to look upwards and leads us from this world to another."

Today, astronomers on Maunakea continue the proud tradition of stargazing begun by our ancestors long ago. Although the technology has changed over time—from naked-eye astronomy to multimillion-dollar telescopes—the quest remains the same: to understand our origin, our place in the cosmos, and perhaps our destiny. PHOTO: RICHARD WAINSCOAT

Skyward

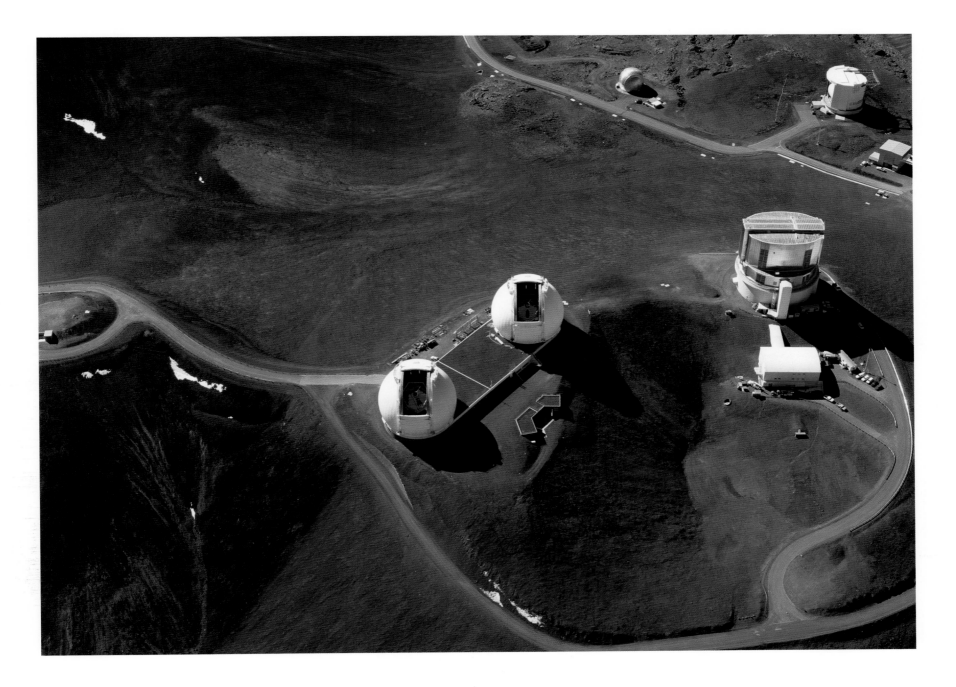

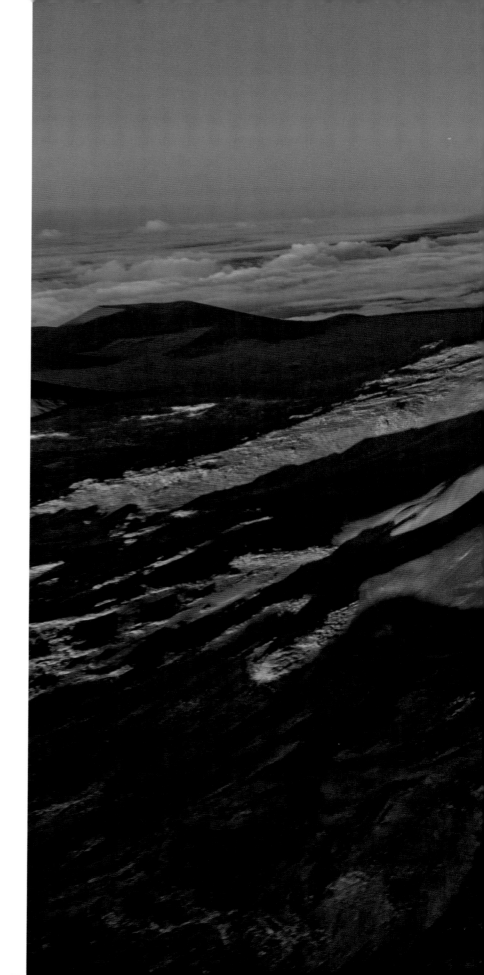

A Bird's-Eye View of the
Maunakea Observatories

The setting sun casts long shadows across Maunakea in this 2014 aerial photograph. Thirteen of the biggest and most sophisticated telescopes ever built stand near the mountain's summit. The same curiosity to find what lies beyond the horizon that inspired the ancient Polynesians to set sail for new lands and first brought them to Hawai'i long ago has inspired astronomers on Maunakea for more than half a century as they explore the cosmic ocean that surrounds our island Earth. PHOTO: W. M. KECK OBSERVATORY AND RIC NOYLE

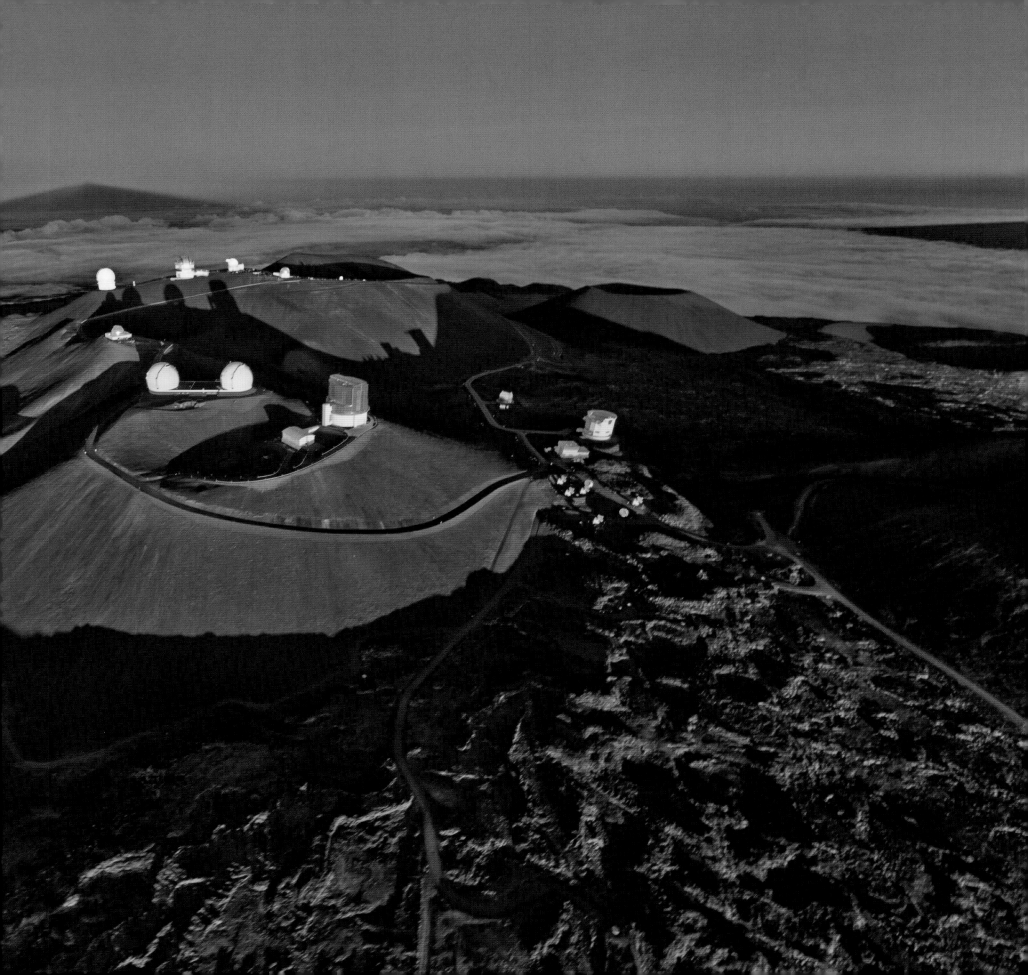

A Multinational Village

Astronomical research is an international endeavor, with more than ten thousand professional astronomers around the world. Maunakea's exquisite observing conditions have made it the location of choice for telescopes from many countries, creating a highly international community on the mountain. On any given day, one might hear French, Japanese, Dutch, Spanish, or other languages spoken on the mountain in addition to English.

Here the Subaru Telescope, owned and operated by the National Astronomical Observatory of Japan, overlooks the Submillimeter Array, a joint venture between the Smithsonian Astrophysical Observatory in the United States and the Academia Sinica Institute of Astronomy in Taiwan.

Observatory partnerships evolve with time as the scientific needs and resources of different communities change. Gemini Observatory, for example, opened in 1999 as an international collaboration between seven countries: Argentina, Australia, Brazil, Canada, Chile, the United Kingdom, and the United States. Telescope time is allocated to astronomers in each partner country in accordance with the amount of financial support provided by each nation. In 2012, however, the UK withdrew from the Gemini partnership, citing funding issues. Australia has also announced that it cannot continue its commitment beyond 2015. New partnerships may be pursued in the coming years as Gemini and other observatories strive to serve the changing needs of their astronomical communities. PHOTO: GLEN PETITPAS

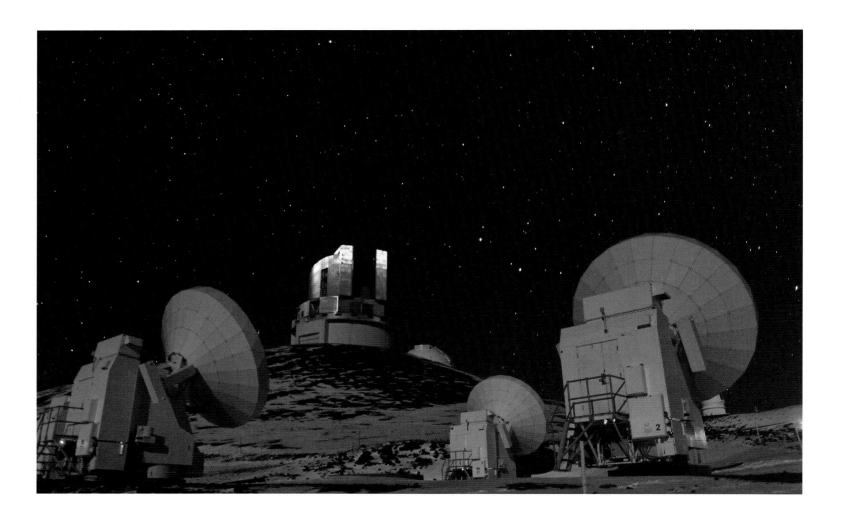

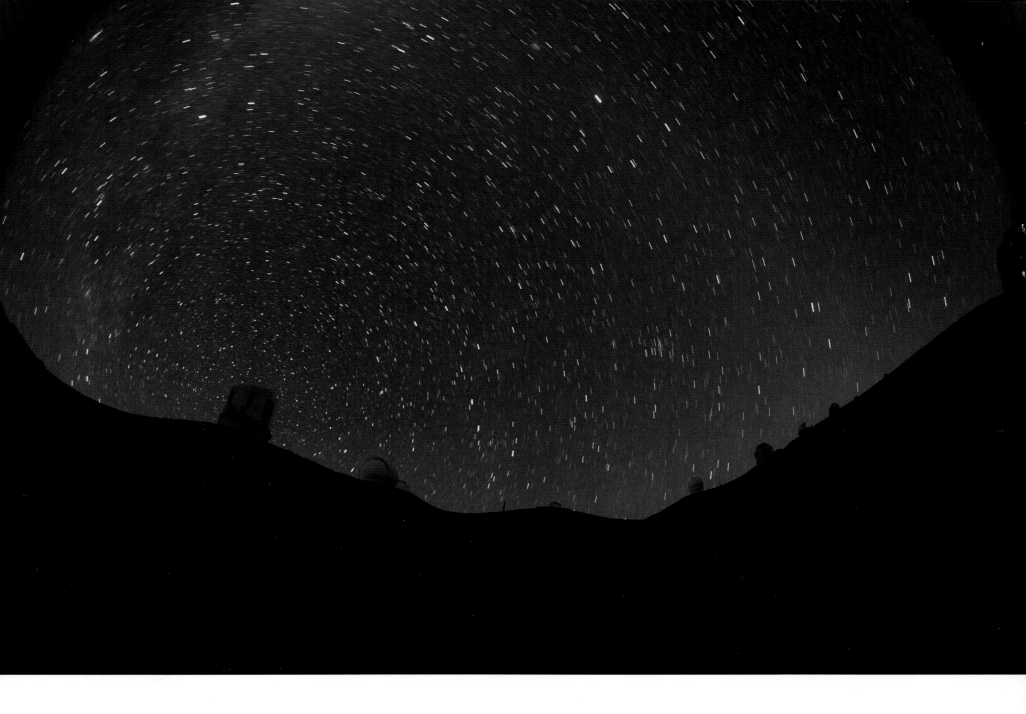

Nine telescopes of different shapes and sizes can be seen in this nighttime photo of Maunakea's summit ridge, providing a snapshot of decades of astronomical development on the mountain. From its humble beginning with two 24-inch telescopes in 1968, Maunakea's reputation as a superb astronomical site quickly attracted international interest. Today there are more major telescopes on Maunakea than any other mountain peak in the world.

PHOTO: SEBASTIAN EGNER AND SUBARU TELESCOPE/NAOJ

Under the Dome
of Night

Beautiful Horizon

This striking panorama shows the University of Hawai'i's 88-inch (2.2-meter) telescope and its neighbors on Maunakea. The wide-angle lens used to obtain this photograph distorts the relative sizes of the different telescopes. PHOTO: 'IMILOA ASTRONOMY CENTER AND HAWAI'I PREPARATORY ENERGY LAB; PHOTO PRODUCED BY MARIKO THORBECKE, BO BLECKEL, HANNAH TWIGG-SMITH, AND PHONG HOANG

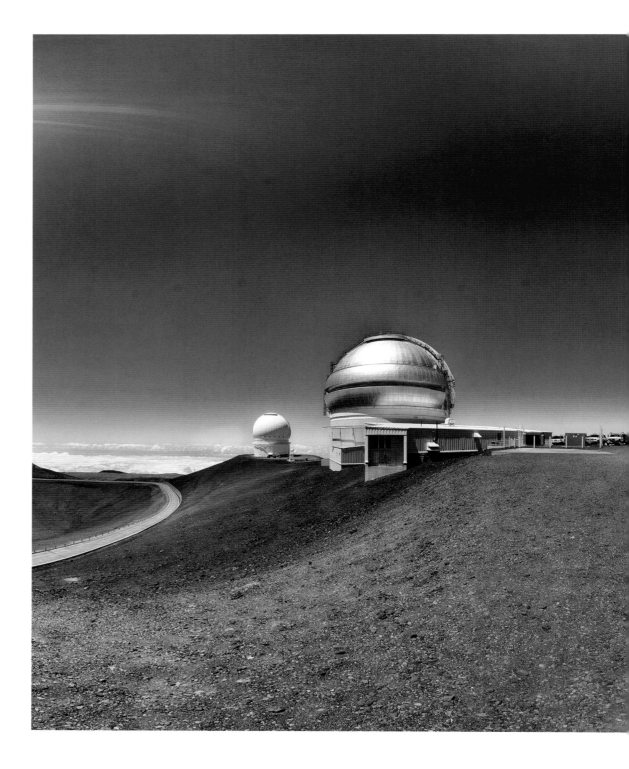

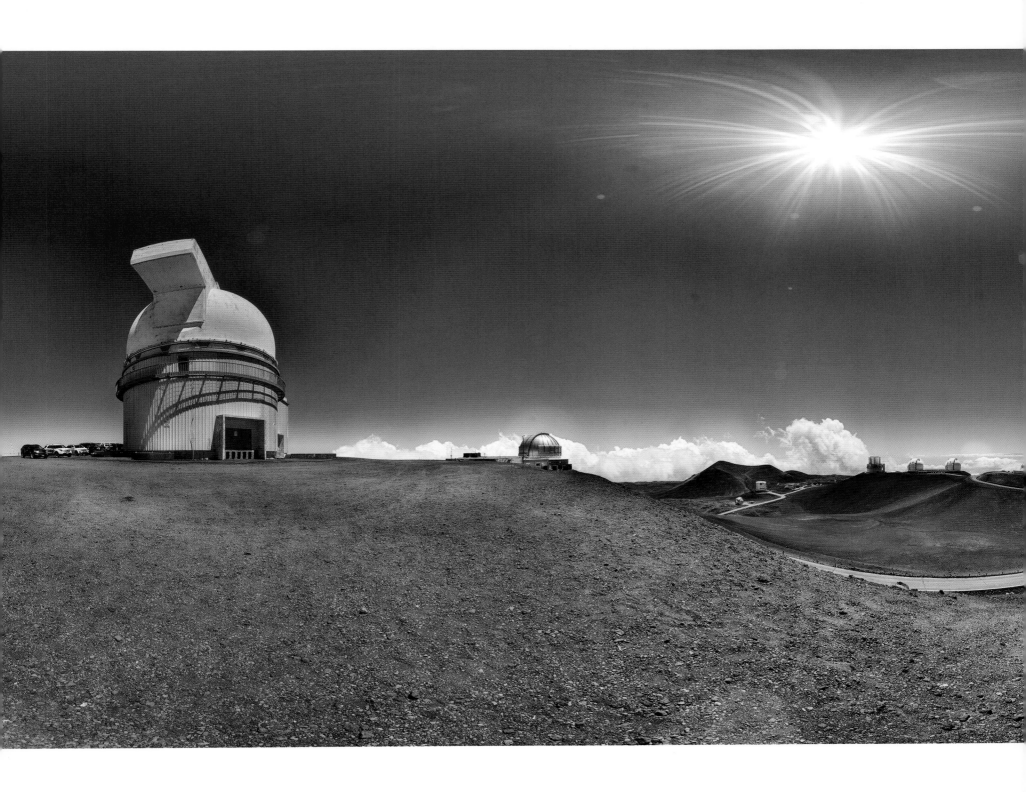

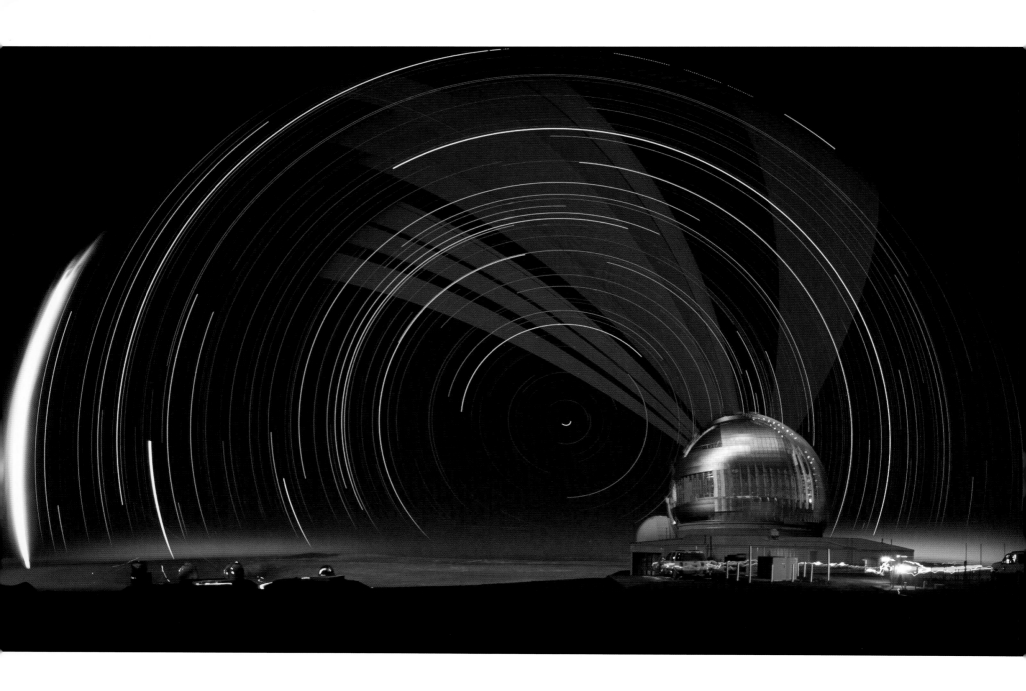

Sunset to Sunrise

This remarkable photograph captures an entire night on Maunakea. Stars appear to arc across the sky as our planet spins, leaving luminous trails in this image, which was made by combining many thirty-second exposures taken throughout the night. Only the North Star remains almost stationary, its low elevation a consequence of Hawai'i's low geographical latitude.

Gemini Observatory is in the foreground, with the Canada-France-Hawai'i Telescope's white dome visible behind it. Four other telescopes—Subaru, Keck I, Keck II, and the NASA IRTF—can also be seen in the distance. Gemini's laser guide star system was used throughout the night to create artificial stars high overhead, the laser's orange beam sweeping across large swaths of sky as the telescope pointed in different directions over the hours. PHOTO: GEMINI OBSERVATORY/AURA AND JOY POLLARD

No other astronomical site in the world boasts such a diverse collection of telescopes from different nations and organizations as Maunakea. Naturally, there's some competition between observatories to make discoveries first, but it's a friendly rivalry, and observatories often work together in the pursuit of scientific knowledge.

In 2014, for example, an international team of astronomers used the Subaru Telescope to take images of a small galaxy that's being shredded by the intense gravity of a larger neighbor, and follow-up observations at Keck Observatory were used to measure the motions of stars as the smaller galaxy is torn apart. That same year, Keck Observatory and Gemini Observatory teamed up to discover the first Earth-sized planet in the habitable zone of another star.

In that same spirit, some Maunakea observatories have exchanged nights of observing time to provide access to new instruments and new observing capabilities for astronomers in their communities.

This photograph shows the laser guide star systems of four telescopes in action simultaneously on Maunakea—Gemini North (out of view), Keck II, Keck I, and Subaru—while each observed regions near the center of the Milky Way. The scientific goals and instruments used to analyze the collected light are different for each telescope, but the quest remains the same—to learn more about the universe and to share the acquired knowledge with the world. PHOTO: JASON CHU

Scientific Synergy

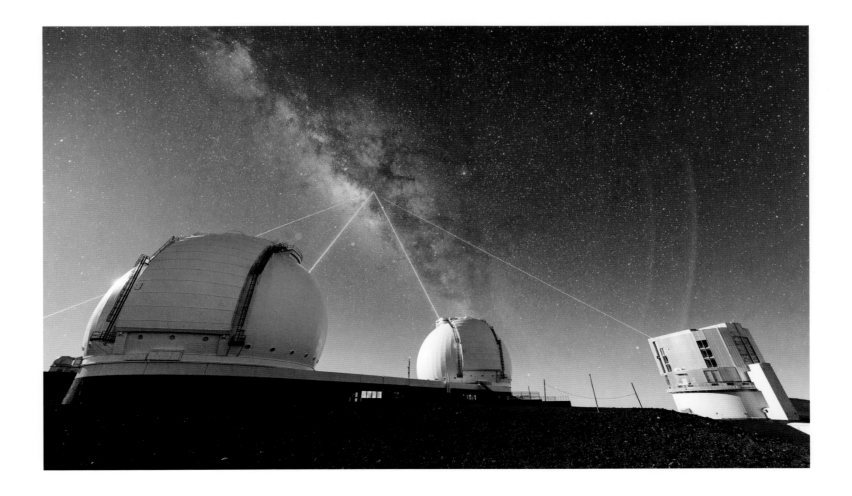

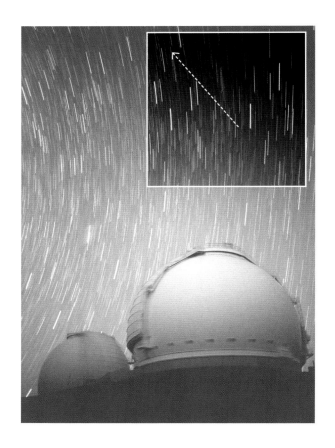

This remarkable photo shows the Hubble Space Telescope (HST) passing over Maunakea in 1996. The orbiting telescope is visible as a faint straight streak of light (track indicated by arrow above) among the star trails near the top of this long exposure. HST circles the Earth at an altitude of about 350 miles (563 kilometers) and completes one trip around our planet every ninety-seven minutes.

Hubble is a relatively small telescope by today's standards; its 7.9-foot (2.4-meter) mirror is dwarfed by Maunakea's giants, such as the 33-foot (10-meter) mirrors of the Keck I and Keck II telescopes seen here. What it lacks in size, however, it makes up for in location. From its vantage point above the Earth's atmosphere, HST is able to obtain razor-sharp images of the cosmos without the atmospheric distortions that affect observations made with earthbound telescopes.

Telescopes in space and on the ground contribute to advancing our knowledge of the universe in different and often complementary ways. For example, Keck Observatory and the Hubble Space Telescope both played essential roles in the discovery of the accelerating expansion of the universe, research that led to astronomers Saul Perlmutter, Adam Riess, and Brian Schmidt being awarded the 2011 Nobel Prize in Physics. HST images were used to identify supernova candidates—the explosive deaths of massive stars that can be seen at great distances—and follow-up spectroscopic observations made at Keck Observatory and elsewhere revealed the characteristics of each supernova and the distances of their host galaxies, information that led to the conclusion that some mysterious force—called dark energy—is causing the universe to speed up as it expands. PHOTO: RICHARD WAINSCOAT

Hubble over Maunakea

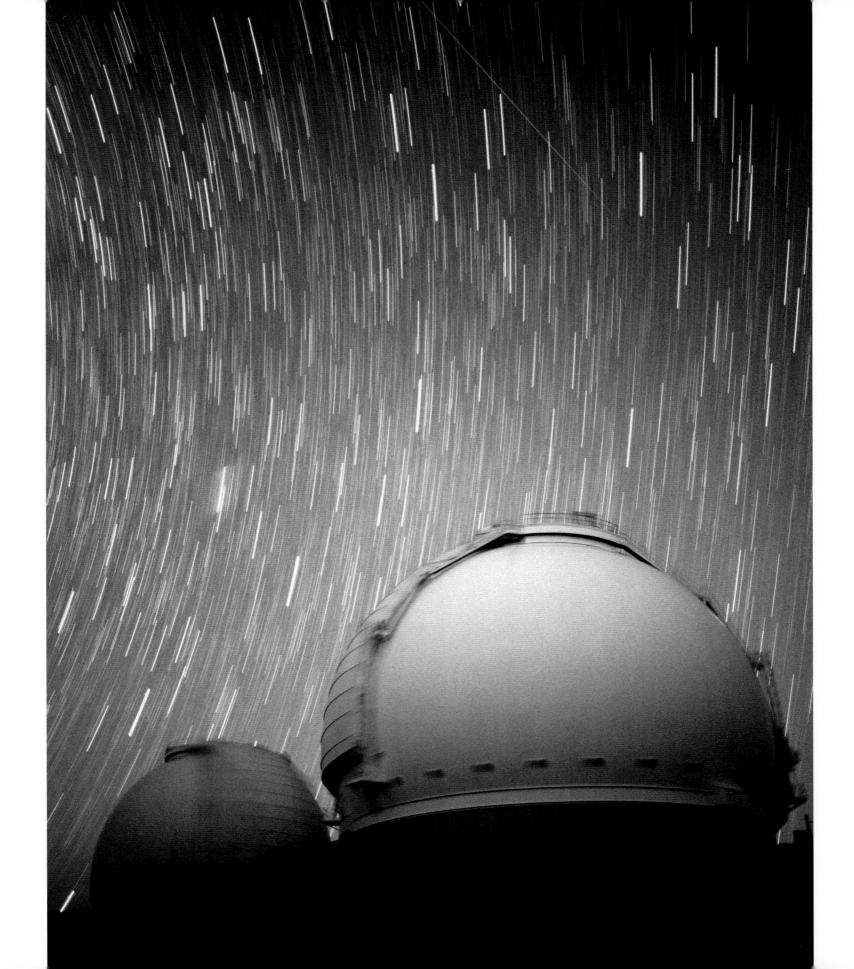

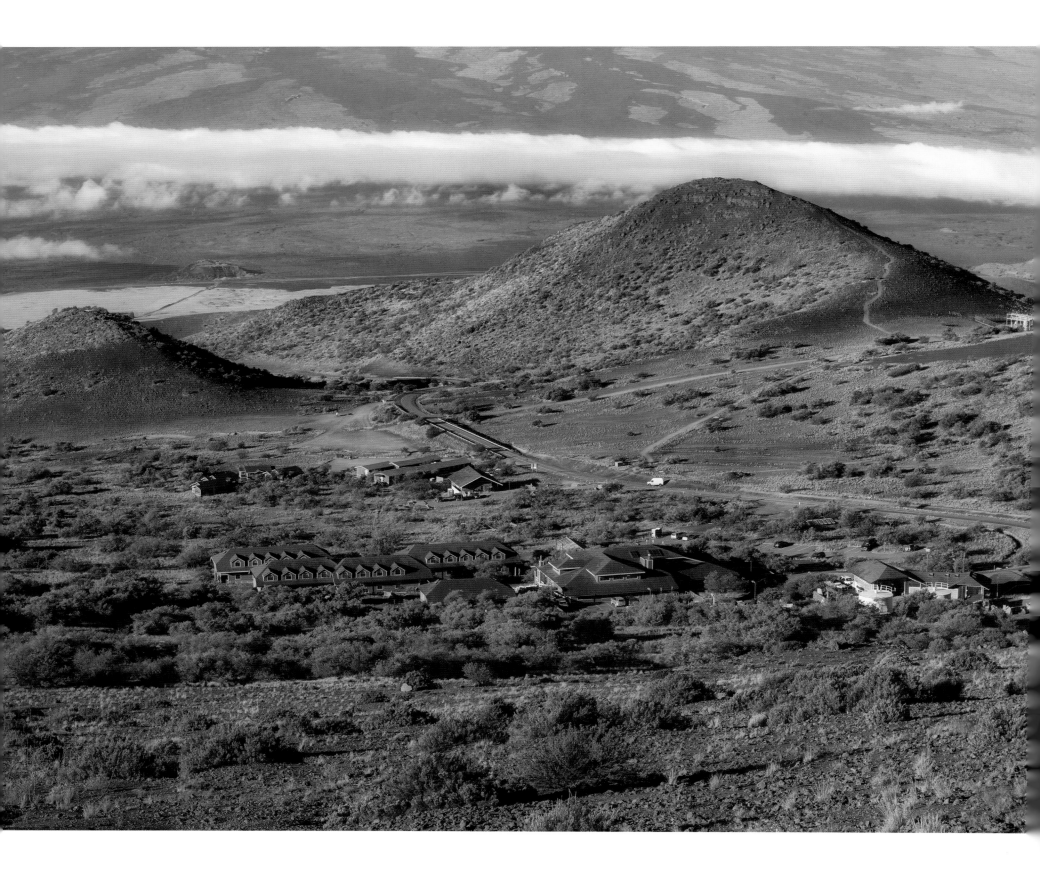

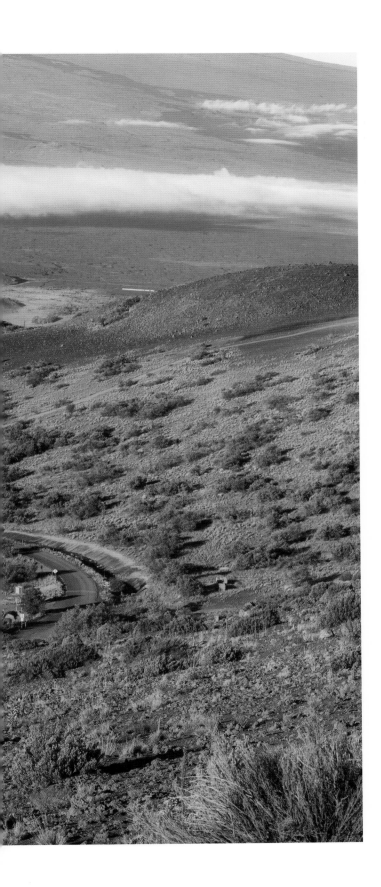

Hale Pōhaku—The Stone House

Astronomers, engineers, students, rangers, and Maunakea Visitor Information Station staff work, sleep, eat, and socialize in Hale Pōhaku, the "stone house." Located at an elevation of 9,300 feet (2,800 meters), Hale Pōhaku began as a few small cabins built in the 1930s for hunters and adventurers. It housed work crews during construction of the first telescopes on Maunakea in the 1970s and became a permanent residence for observatory staff and visiting astronomers in 1983.

Hale Pōhaku can provide accommodations for up to seventy-two people. However, as remote operation of the telescopes on Maunakea has increased over the years or shifted to service mode in which observatory staff astronomers do observations on behalf of other astronomers around the world, the number of people who need to stay overnight at Hale Pōhaku has decreased, so that today occupancy averages about twenty-five people per night. The number will undoubtedly increase again in the near future when construction of the new Thirty Meter Telescope begins. PHOTO: MICHAEL WEST

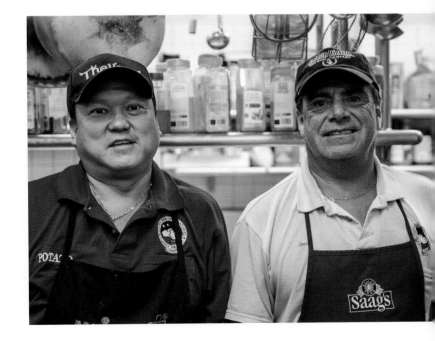

Feeding Body and Soul

Maunakea has nourished the human spirit of exploration for half a century. But bodies need nourishment too.

More than forty thousand meals are served each year at Hale Pōhaku for astronomers, engineers, rangers, and other staff. Another five thousand lunches are delivered to workers on the mountain's summit yearly. This is not an easy task, as food and water must be delivered regularly to this remote site. More than 2 million gallons of water are needed yearly—for cooking, cleaning, showering, and other needs—requiring trucks to deliver water to Hale Pōhaku several times per week.

Jason Hashimoto and Steven Garcia have each worked in the kitchen at Hale Pōhaku for decades. Without them, and all of the dedicated support staff on Maunakea, the many scientific discoveries made with the telescopes wouldn't be possible. PHOTO: MICHAEL WEST

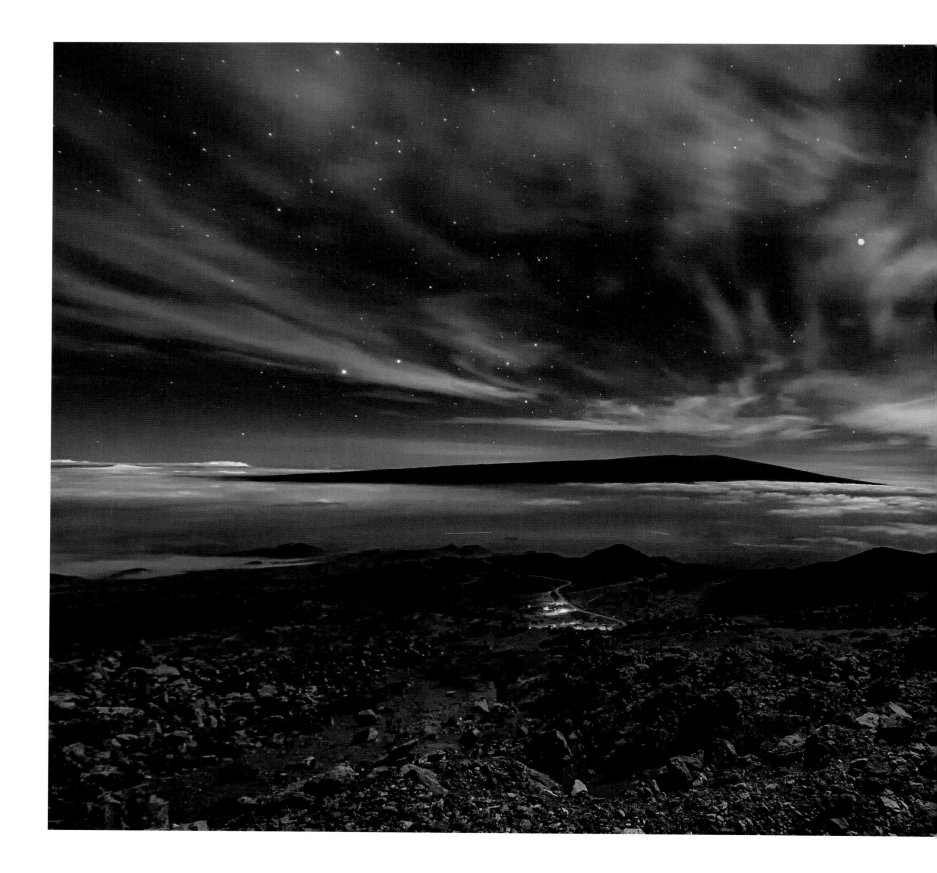

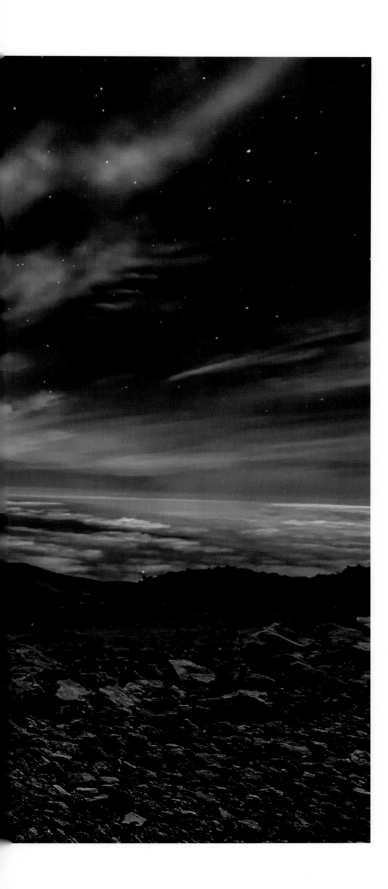

Hale Pōhaku and Maunaloa at Night

As weary astronomers and telescope operators drive down from the mountain after a long night of observing the heavens, Hale Pōhaku's lights are a welcome sight. Maunaloa is visible in the distance, as is the glow of lava flowing from Kīlauea volcano. PHOTO: MICHAEL WEST

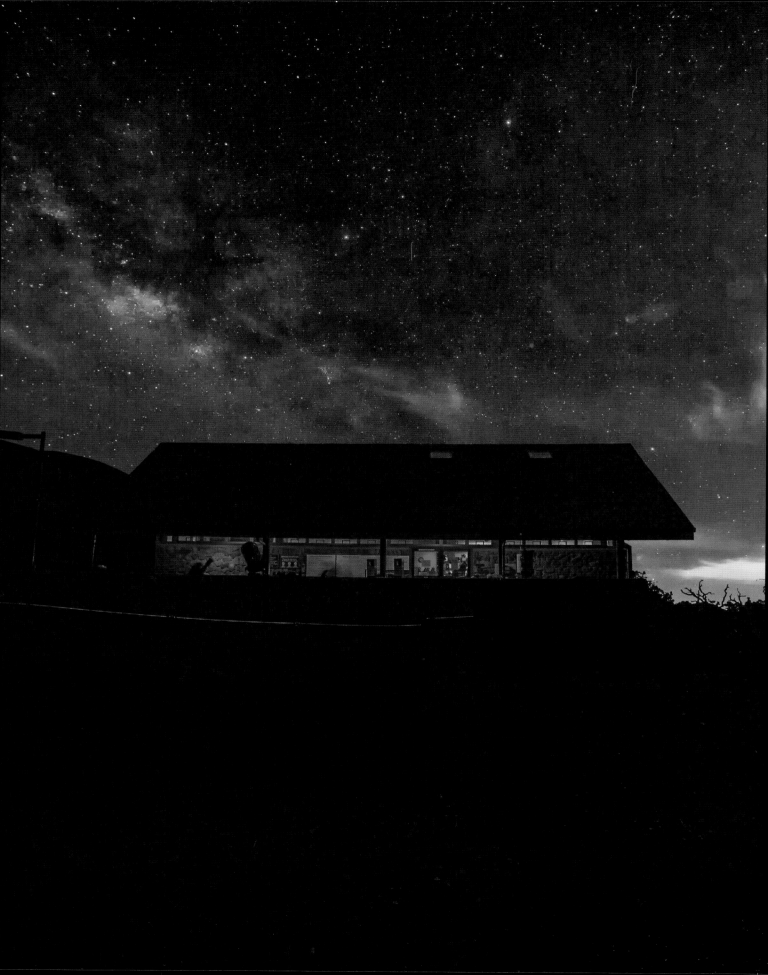

Heaven Above

One of the Big Island's most popular tourist destinations, the Maunakea Visitor Information Station receives more than two hundred thousand visitors a year. Staffed mostly by volunteers, it offers popular lectures on astronomy and Hawaiian culture, public stargazing with telescopes, and hot chocolate or ramen to help visitors stay warm in the cold night air.

The visitor center is located about halfway up the mountain, at an elevation of 9,000 feet (2,740 meters). Anyone intending to go to the summit of Maunakea is strongly encouraged to spend at least thirty minutes at the visitor center to acclimatize to the thin air before proceeding further. Many people don't realize, however, that more stars are visible from the visitor center than from the top of Maunakea, as the reduced oxygen content at the highest elevations diminishes visual acuity. PHOTO: MICHAEL WEST

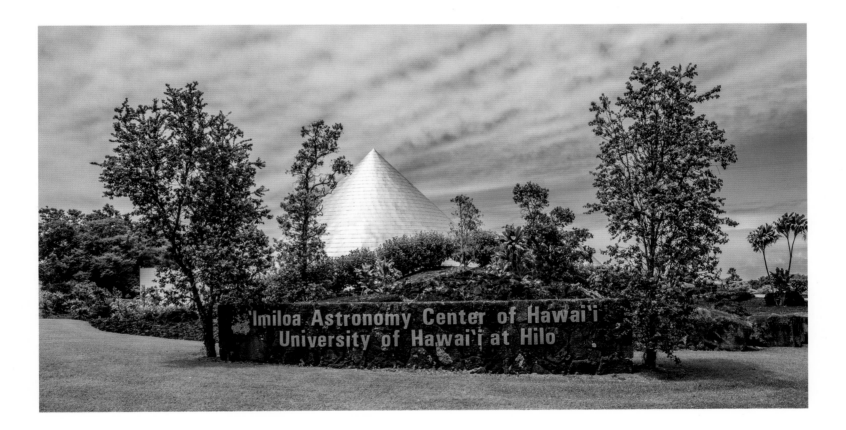

Science and culture come together at the ʻImiloa Astronomy Center of Hawaiʻi. The $28 million NASA–funded center, located on the campus of the University of Hawaiʻi at Hilo, has welcomed almost a million visitors since opening its doors in February 2006.

Created with participation from the astronomical community in Hawaiʻi, the University of Hawaiʻi, and the Native Hawaiian community, the center aims to inspire visitors of all ages to reach for the stars and to embrace the cultural heritage of the Islands. Within its 40,000 square feet (3,700 square meters) of space, it features exhibits, shows—including a 120-seat planetarium—lectures, and other activities that reflect Maunakea's importance to Hawaiians and astronomers alike. PHOTO: MICHAEL WEST

The ʻImiloa Astronomy Center of Hawaiʻi

In 1928, American astronomer George Ellery Hale wrote an article for *Harpers Monthly* magazine titled "The Possibilities of Large Telescopes." Hale, who had founded three observatories—Yerkes, Mount Wilson, and Palomar—and built the world's largest telescope four times, wrote, "While the gain since Galileo's time seems enormous, the possibilities go far beyond. Starlight is falling on every square mile of the earth's surface, and the best we can do at present is to gather up and concentrate the rays that strike an area one hundred inches in diameter."

Today's telescopes would surely have impressed Hale. Maunakea is home to the largest optical and infrared telescopes in the world—Keck I and Keck II—and there are currently more square inches (square centimeters) of mirror glass on the mountain than anywhere else on Earth. But astronomers' ability to explore the universe, to see better and farther, is limited by telescope size. Plans are already underway for a new generation of extremely large telescopes that will push the limits of technology and enable whole new areas of astronomical research.

Maunakea has been chosen as the site of the ambitious Thirty Meter Telescope (TMT). Conceived as a first-of-its-kind international collaboration between public and private partners in Canada, China, India, Japan, and the United States—with additional partners still possible—the TMT will revolutionize astronomy.

As its name implies, the TMT will use a mirror 30 meters (98.4 feet) in diameter to collect nearly ten times as much light as the Keck Observatory telescopes. This enormous increase in light gathering power will not only allow fainter objects to be seen clearer than ever before, it will also provide a nearly tenfold increase in resolving power, the ability to see fine details. TMT's giant mirror will be made of 492 individual segments, each 4.75 feet (1.45 meters) in diameter. A built-in adaptive optics system will help to eliminate image blur that occurs when light passes through Earth's atmosphere, yielding exquisitely sharp images that will surpass anything obtainable with the Hubble Space Telescope. The telescope's dome will also have a unique design, rotating in two different planes.

Groundbreaking for the TMT took place on October 7, 2014. Construction is expected to take nearly a decade, with the first scientific observations likely in the early 2020s.

IMAGE: TMT OBSERVATORY CORPORATION

The Thirty Meter Telescope

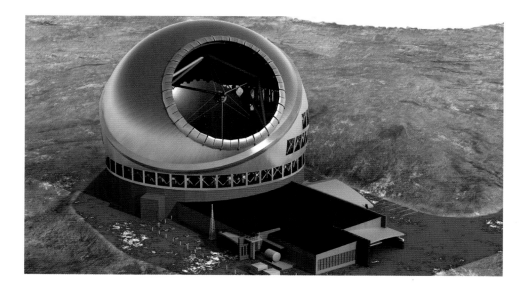

Maunakea was chosen as the future home of the Thirty Meter Telescope (TMT) after an exhaustive evaluation of possible sites in Chile, Mexico, and the United States.

In his 2014 State of the State address, Hawai'i governor Neil Abercrombie said, "Maunakea is Hawai'i's gift to the world—the best place on the planet to observe the universe. It is without peer." The governor noted that construction of the TMT "will solidify Hawai'i's position as the world's premier astronomy center," and he emphasized the many benefits that astronomy brings to the Islands by providing jobs, education, and inspiration. TMT, for example, will invest $1 million every year in education to help Hawai'i's children reach for the stars. "Our state must support and ensure that this tremendous opportunity comes to fruition," said the governor.

This artist's impression shows TMT's planned location on a barren plateau along Maunakea's northern slope, slightly below the other telescopes along the mountain's summit ridge. This particular location was chosen for both scientific and sociological reasons, including the fact that it will be invisible from most of the island of Hawai'i and will have less potential impact on cultural and archaeological sites of importance to some Native Hawaiians. IMAGE: TMT OBSERVATORY CORPORATION

Future Site of the Thirty Meter Telescope

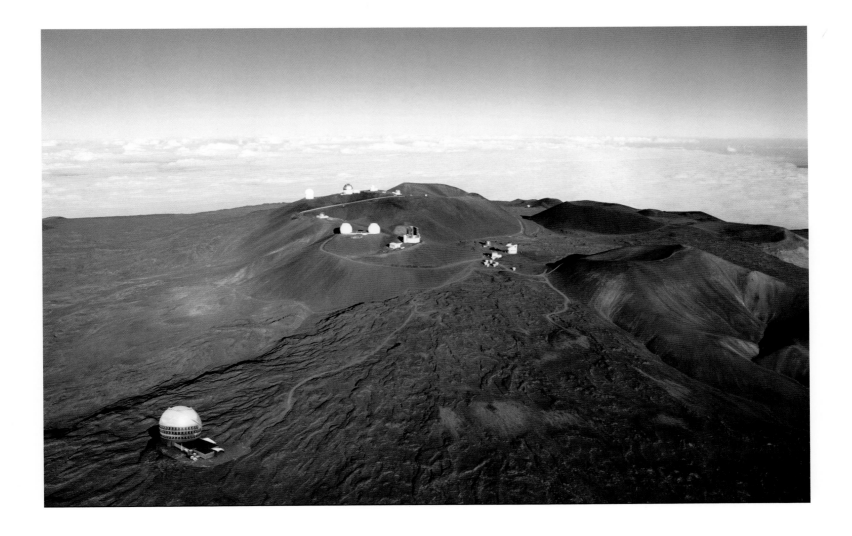

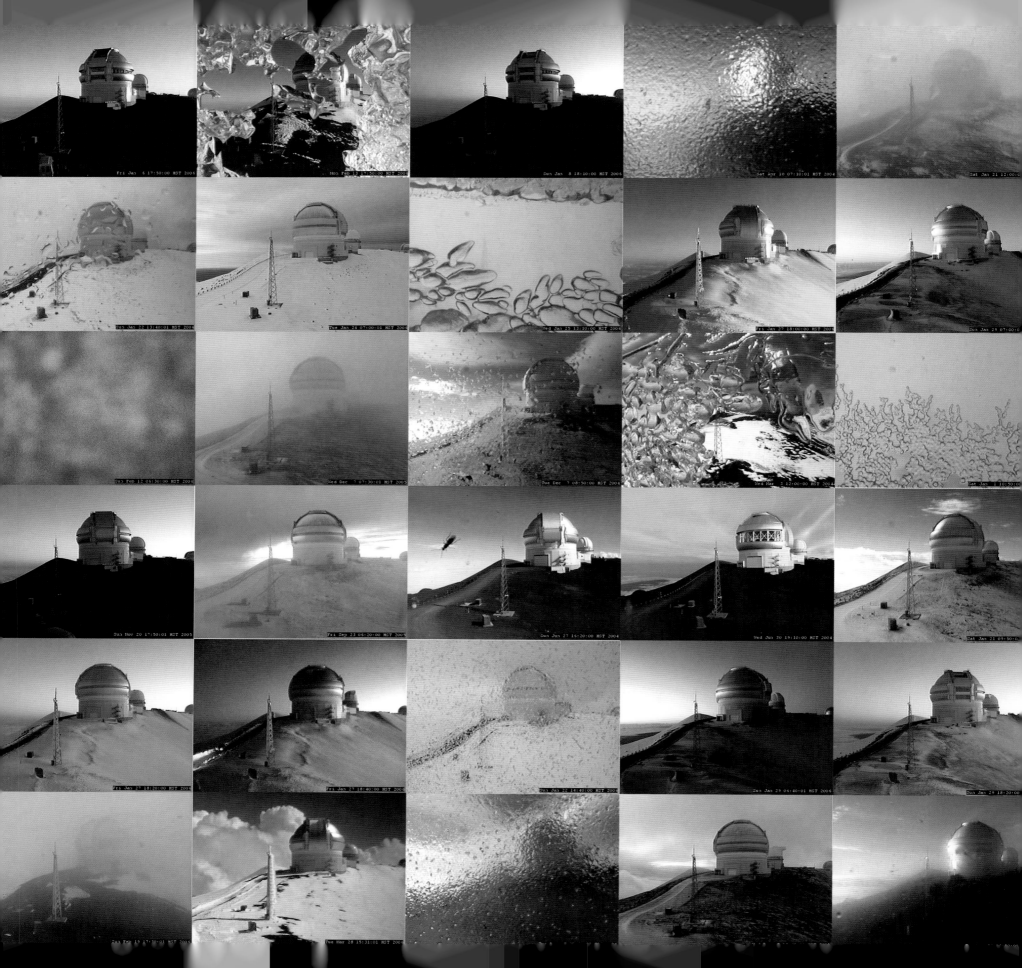

The island of Hawai'i holds the records for hottest and coldest temperatures ever recorded in the state. The temperature reached a sweltering 100 degrees Fahrenheit (38 degrees Celsius) in the town of Pāhala on April 27, 1931. The coldest temperature ever recorded in Hawai'i was—not surprisingly—on the summit of Maunakea, where it fell to a frigid 12 degrees Fahrenheit (–11 degrees Celsius) on May 17, 1979.

Although skies are clear most nights on Maunakea, the weather is never dull and often unpredictable. This montage of images, compiled over the course of a year from the Canada-France-Hawai'i Telescope's outdoor webcam as it looks towards the Gemini North Telescope, shows some of the diversity of conditions encountered on Maunakea.

The Maunakea Weather Center, a cooperative effort of the Institute for Astronomy and Department of Meteorology at the University of Hawai'i at Mānoa, provides detailed weather forecasts for the mountain's summit, including likely observing conditions each night. PHOTOS: MICHAEL WEST AND CANADA-FRANCE-HAWAI'I TELESCOPE

Wicked Weather

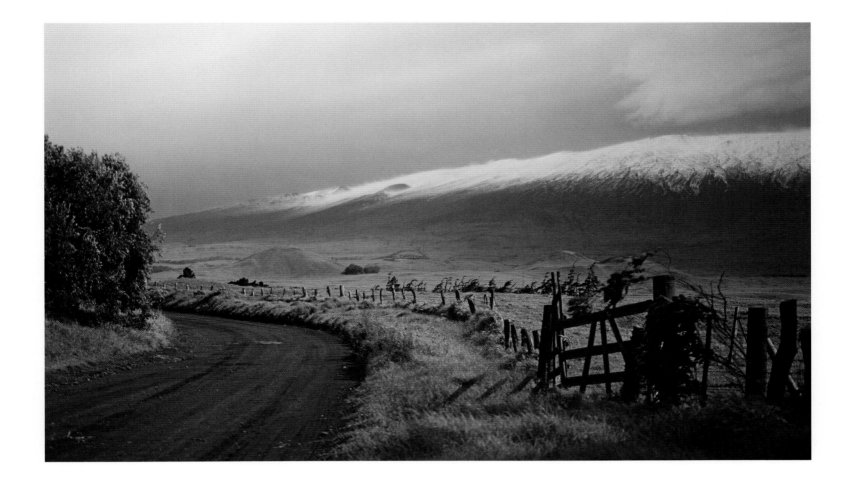

Snowcapped Maunakea

Maunakea lies under a blanket of freshly fallen snow as seen from the bucolic town of Waimea, home to the headquarters of the W. M. Keck Observatory and the Canada-France-Hawai'i Telescope. PHOTO: BARBARA SCHAEFER

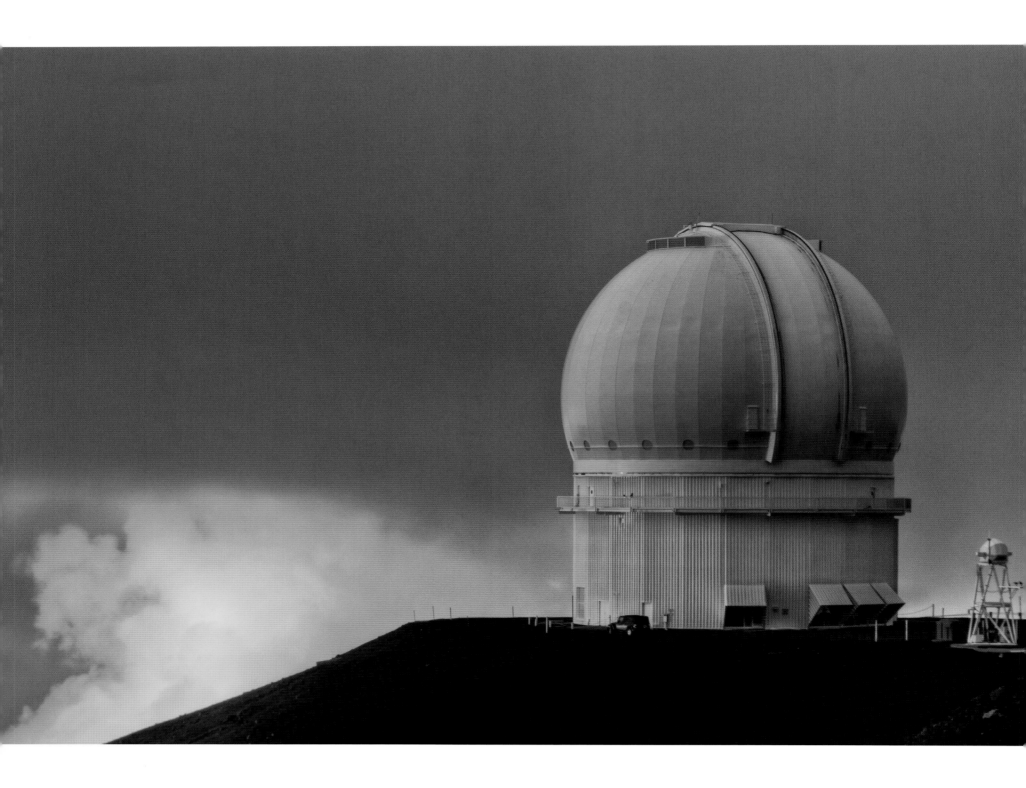

Tempestuous Twilight

Sunsets come in all colors on Maunakea. This photograph shows the Canada-France-Hawai'i Telescope bathed in a fiery glow from sunlit clouds below. PHOTO: TOM KERR AND JOINT ASTRONOMY CENTRE

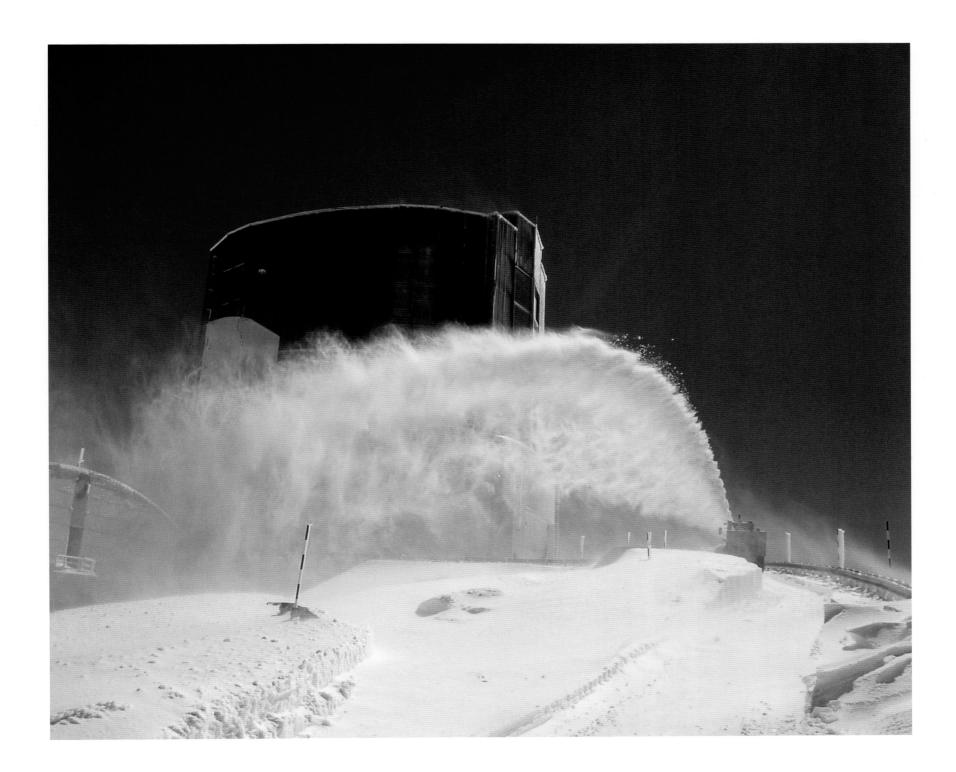

Sometimes getting to work can be a challenge on Maunakea. Here a snowblower clears freshly fallen snow from around the Subaru Telescope. PHOTO: SAEKO HAYASHI AND SUBARU TELESCOPE/NAOJ

Snowy Summit

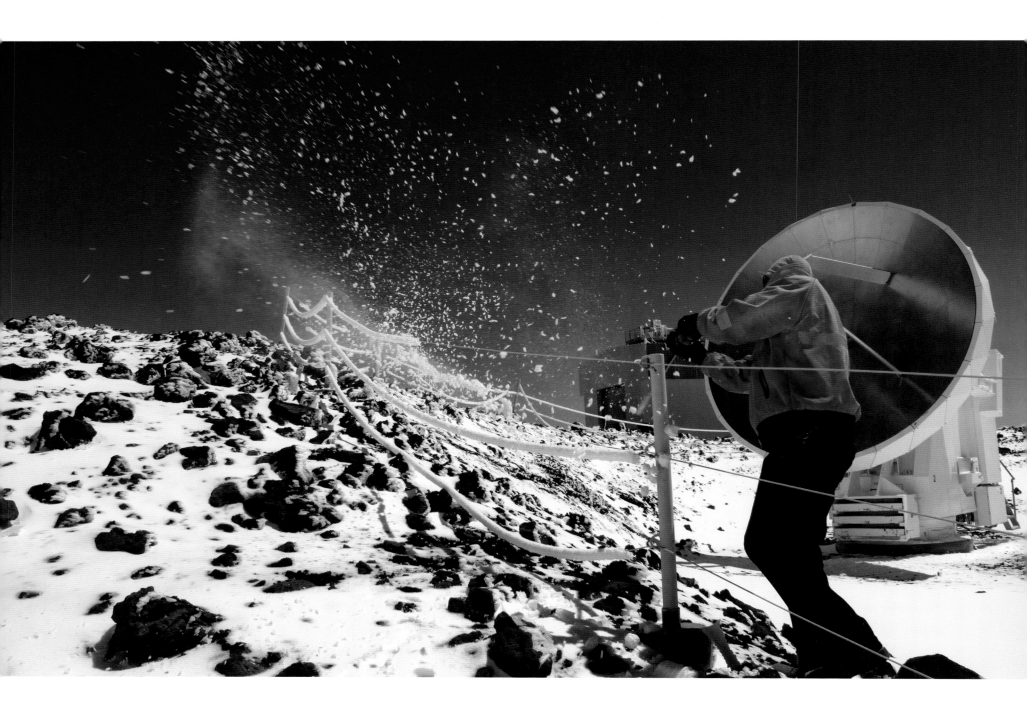

Breaking the Ice on Maunakea

Chunks of ice go flying near the Submillimeter Array on a cold but sunny day following a winter storm. PHOTO: JENNIFER MILLER

A cloudy night on Maunakea is bad news for astronomers on any telescope. But even a clear one can sometimes pose problems at submillimeter observatories.

Water vapor in the atmosphere absorbs submillimeter light like a sponge absorbs water, preventing it from reaching the ground. To a submillimeter telescope, high humidity is like clouds: the moister the air, the dimmer the view.

Maunakea's dry air is one of the reasons why the Submillimeter Array, Caltech Submillimeter Observatory, and James Clerk Maxwell Telescope were all constructed there. However, weather systems that pass over Maunakea occasionally bring moist skies that make it difficult or impossible for submillimeter telescopes to see the heavens.

PHOTO: GLEN PETITPAS

A Moist Sky

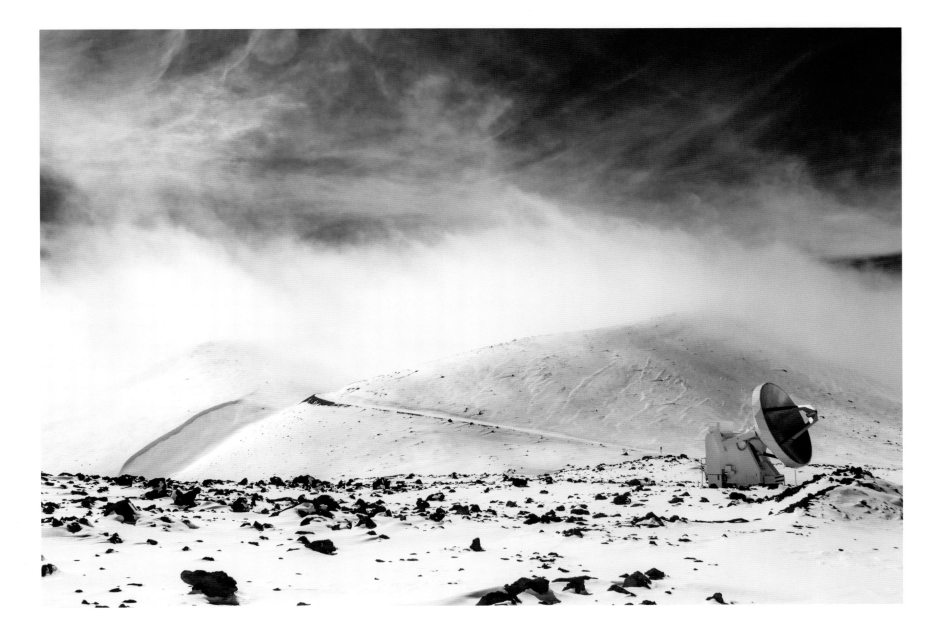

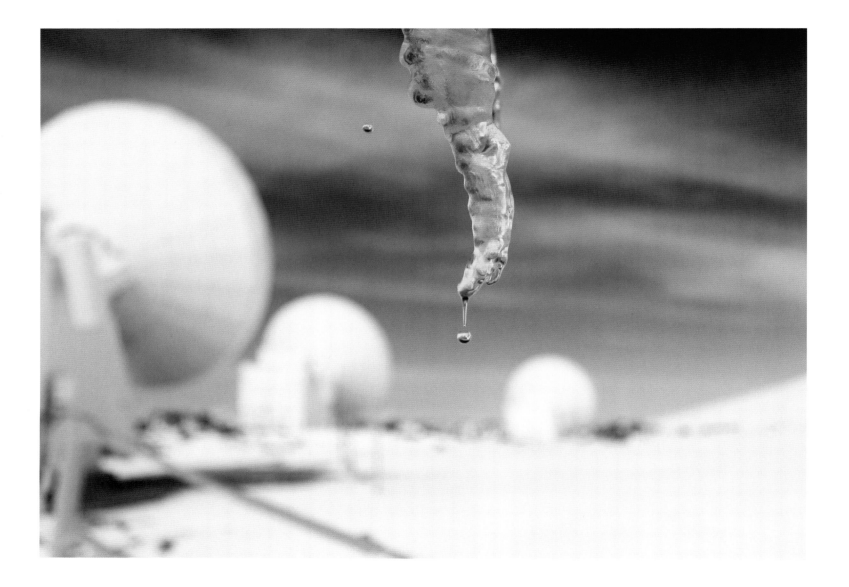

After the Storm

An icicle hangs from one of the eight dishes comprising the Submillimeter Array.
PHOTO: GLEN PETITPAS

Astronomer Richard Wainscoat captured this unique 1991 image of the Canada-France-Hawai'i Telescope bathed in the light of the full moon. The sky's unusual pink glow was created by dust from the eruption of Mount Pinatubo, a volcano on the island of Luzon in the Philippines, together with scattering of moonlight by Earth's atmosphere. Pinatubo's outburst, the second largest of the twentieth century, killed nearly a thousand people. The effects of the eruption were felt around the world, even coloring the sky over Maunakea more than 5,000 miles (8,000 kilometers) away. PHOTO: RICHARD WAINSCOAT

Pink Sky at Night

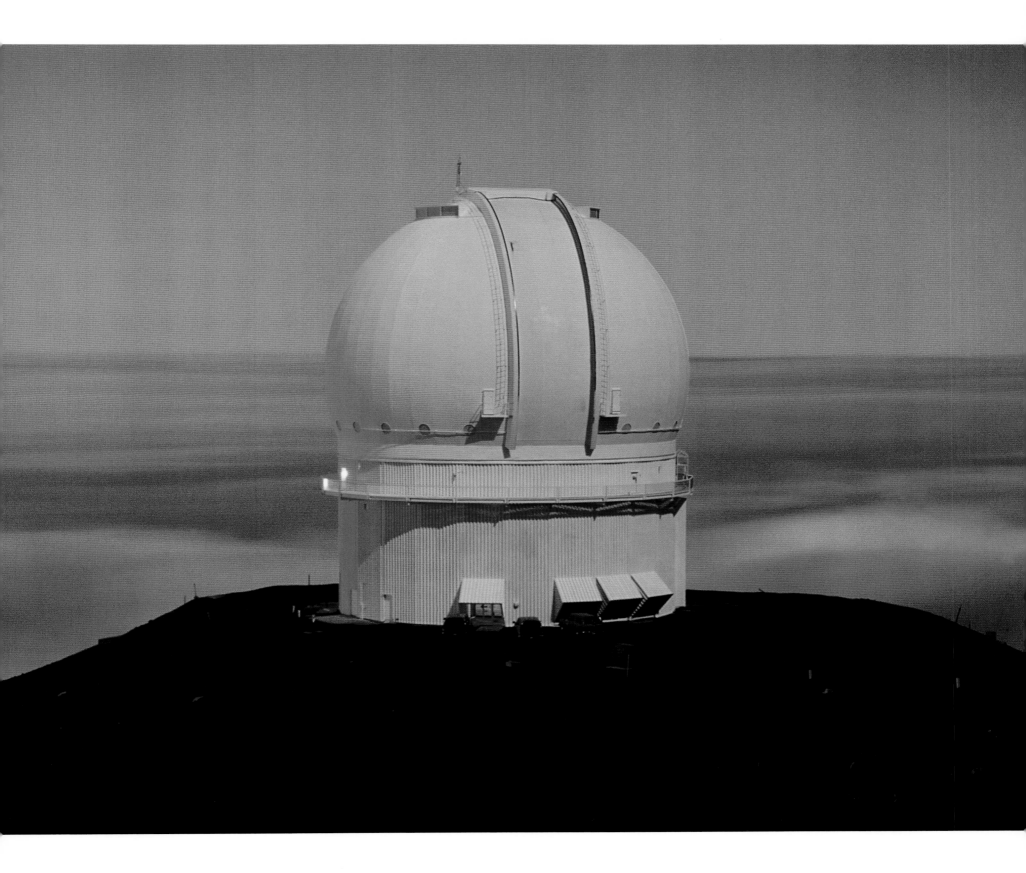

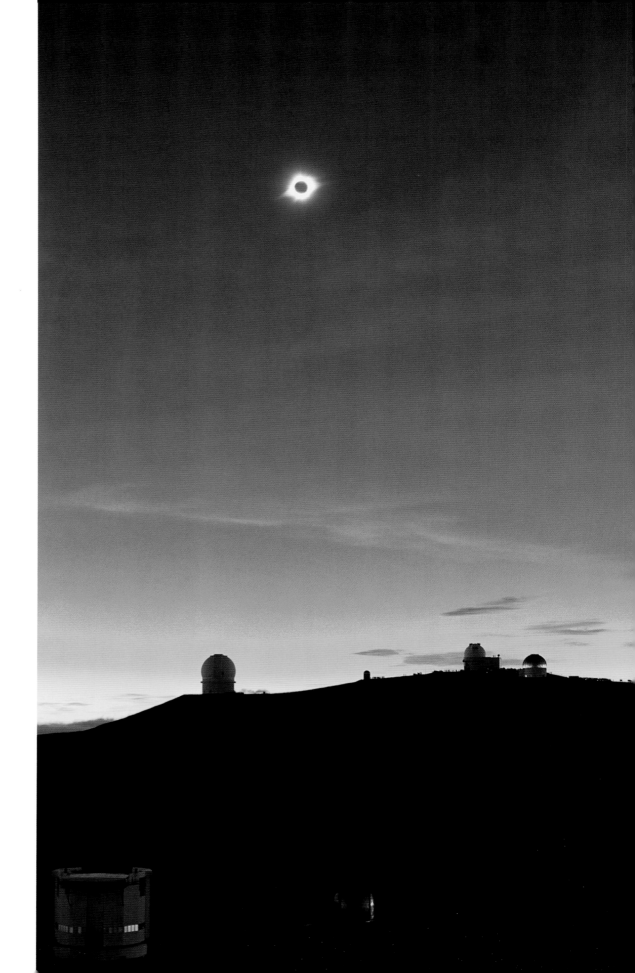

The Vanishing Sun

The day grew dark in the early morning of July 11, 1991, as the Earth, moon, and sun aligned to produce a rare total solar eclipse in the skies over Maunakea. This particular solar eclipse was unusually long, with the moon blotting out the sun for about four minutes over Hawai'i and nearly seven minutes in parts of Mexico and Central America. Much of the island of Hawai'i was covered by thick clouds that day, but the view from the summit of Maunakea was spectacular. There won't be another total solar eclipse of such long duration visible from anywhere on Earth until the year 2132. PHOTO: RICHARD WAINSCOAT

A Rain of Starlight

The stars appear as a rain of light in this time-lapse photograph, the result of our planet's rotation. PHOTO: MICHAEL WEST

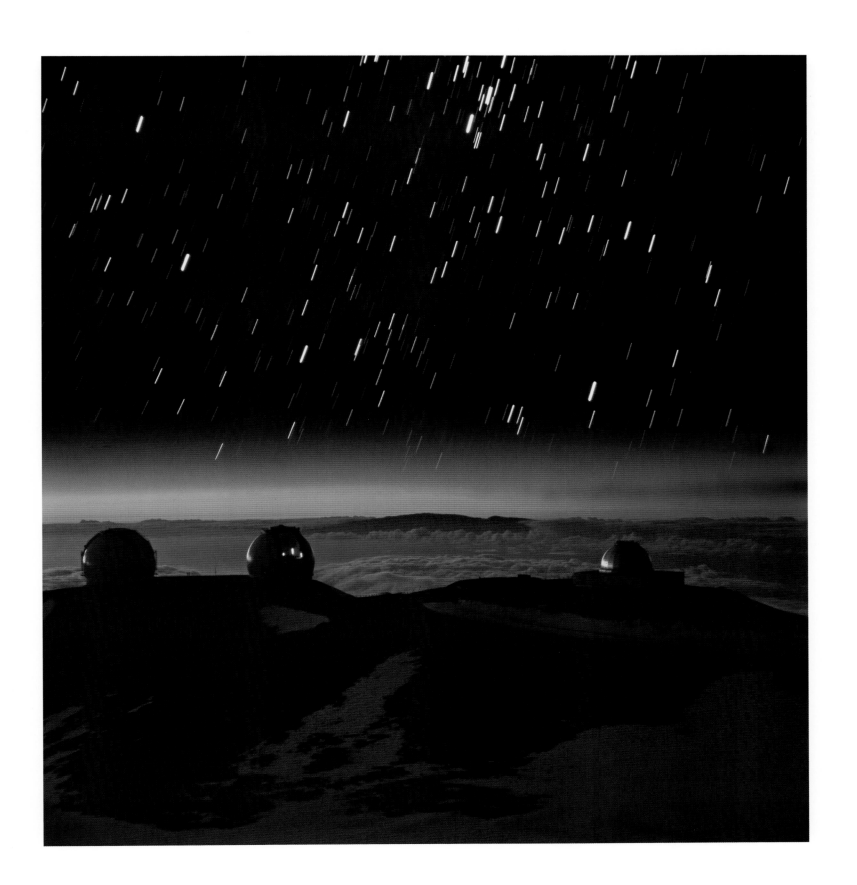

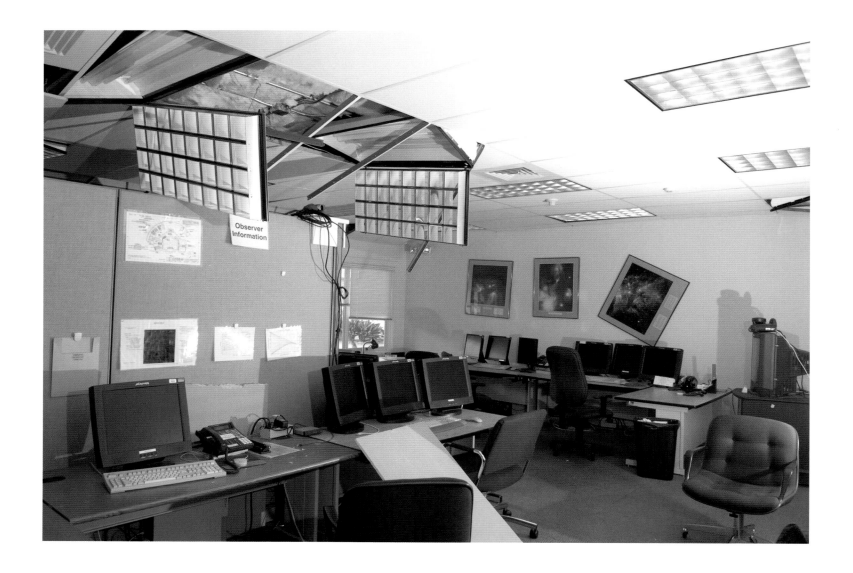

Because the Hawaiian Islands are still geologically active, Maunakea sometimes experiences earthquakes. The most powerful one in twenty years occurred on October 15, 2006, when the Big Island was rocked by two consecutive earthquakes of magnitude 6.7 and 6.0 that struck just a few miles off the Kona Coast.

Special safety features, such as restraints that activate when the ground begins to shake, are designed to prevent or at least minimize damage by safeguarding the delicate mirror and other fragile parts. Nevertheless, the large magnitude of the 2006 quake damaged several telescopes on Maunakea. The jolt moved the entire 300-ton Keck II telescope more than an inch, requiring scientists and engineers to recalibrate its control system—used to aim the telescope—by factoring in its new location. The total cost of repairs to all observatories exceeded $1 million.

This photograph shows damage to the Keck II remote operations control room at Keck Observatory headquarters in Waimea after the 2006 earthquake. PHOTO: W. M. KECK OBSERVATORY AND SARAH ANDERSON

Shake, Rattle, and Roll

Every second of telescope time is precious on Maunakea. Anticipation grows while waiting for the sun to set and another night of cosmic exploration to begin. PHOTO: IRTF AND MIKE CONNELLEY

Anticipation

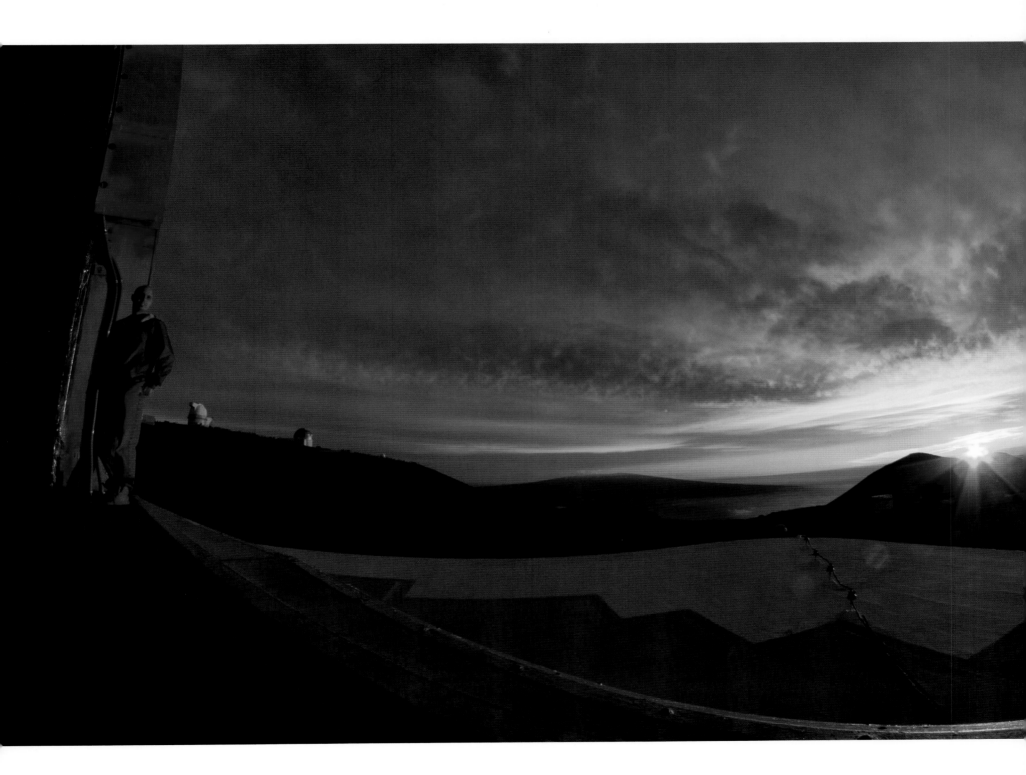

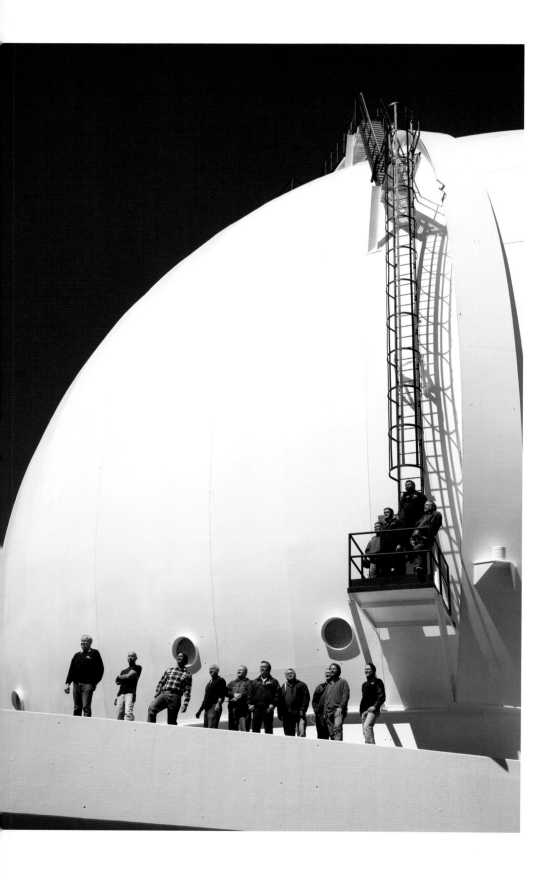

An Observatory's Greatest Asset

Maunakea's location makes it a good place for astronomy, but it's the people who work there that make it great. Each observatory's most valuable asset is its staff.

Tremendous effort goes into keeping the telescopes running in tip-top condition. The stakes are enormous. The cost of operating one of these telescopes translates to nearly a dollar per second, so the last thing anybody wants is to lose precious moments during the night because of technical problems. While astronomers sleep during the day, engineers and technicians work to ensure that everything is ready for the next night. It's challenging but rewarding work that often requires pushing the limits of technology to see the nearest and farthest objects in the universe.

Here the Keck Observatory daytime crew poses for a group shot outside one of the observatory's two domes. PHOTO: W. M. KECK OBSERVATORY AND RICK PETERSON

A Hard Day's Night

Observatories are 24/7 operations, and somebody is always working on Maunakea. For every astronomer observing with one of the telescopes, there are many other people in supporting roles. Telescope operators control the mammoth machines as they explore the night sky. Engineers build and test new instruments for telescopes. Technicians must repair or replace worn parts to keep everything functioning. Administrators organize transportation, deliveries, and accommodations for a constant stream of observatory staff and visiting astronomers. Programmers write software to control the telescopes and to archive the data they collect. Road crews clear snow, ice, and rocks. Rangers keep everyone safe from hazards on Maunakea's roads and summit.

Without the support of this diverse community of dedicated professionals, none of the amazing astronomical discoveries made on Maunakea would be possible. PHOTOS: GEMINI/AURA, SUBARU/NAOJ, W. M. KECK OBSERVATORY, UNIVERSITY OF HAWAI'I INSTITUTE FOR ASTRONOMY, JOINT ASTRONOMY CENTRE, CALTECH SUBMILLIMETER OBSERVATORY, NASA IRTF, AND MIKE CONNELLEY

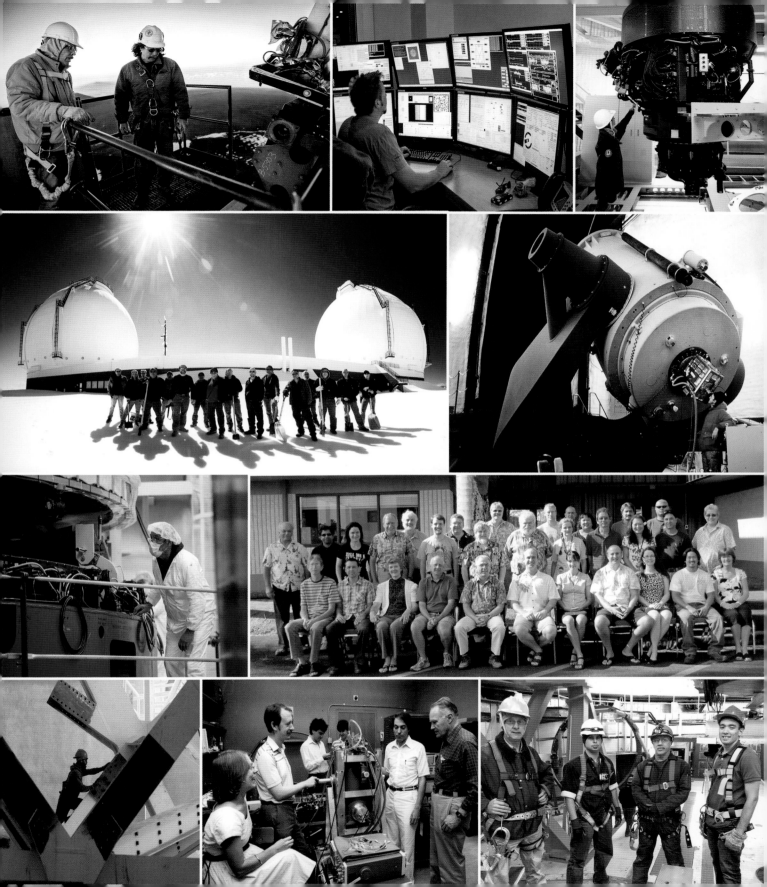

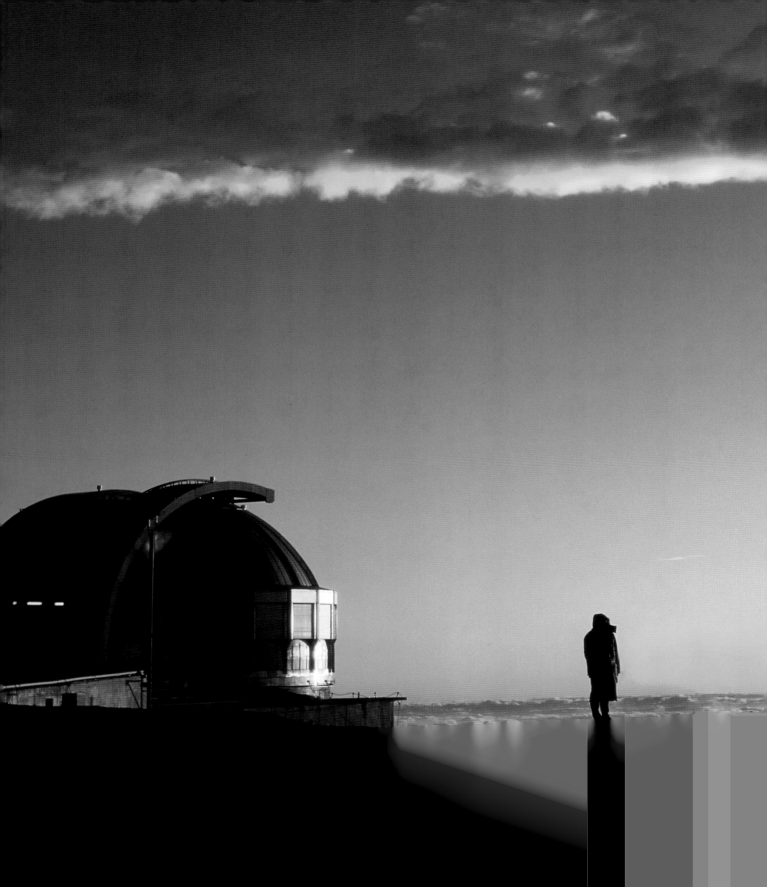

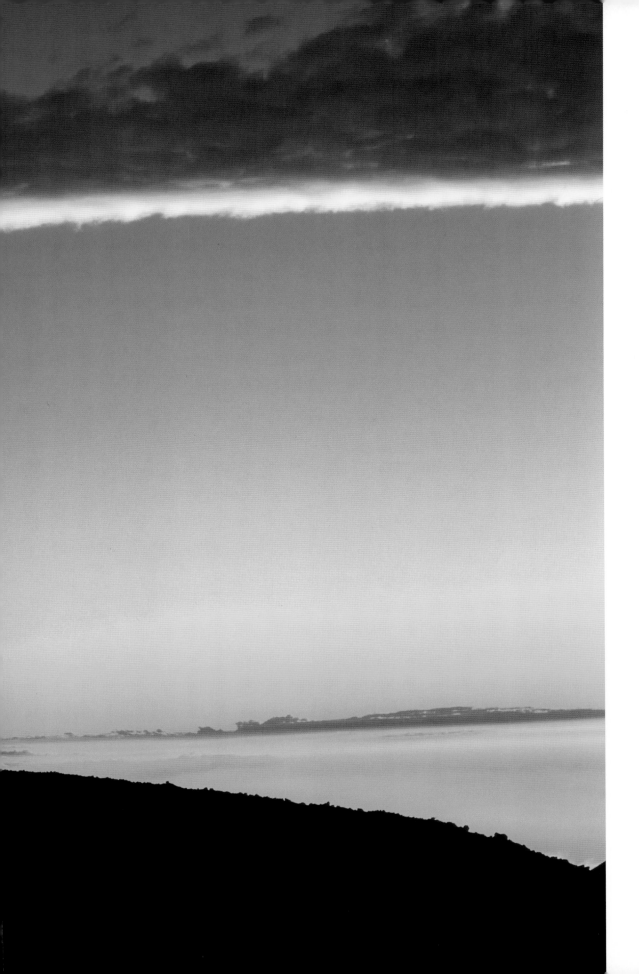

Limited Sight, Limitless Vision

They come to Hawaiʻi not for the sun, sea, or surf, but for the stars. Maunakea's pristine skies have made it a mecca for astronomers from around the world for more than half a century.

To stand on the summit of Maunakea is to inhabit a thin sliver of space wedged between land and sky. Many who have stood here have been moved by the experience, but perhaps none described it better than the famous Scottish botanist David Douglas, for whom the Douglas fir tree is named. "Never, even under a tropic sky, did I behold so many stars," he wrote in his journal in 1834, adding, "Man feels himself as nothing, as if standing on the verge of another world."

PHOTO: MICHAEL WEST

135

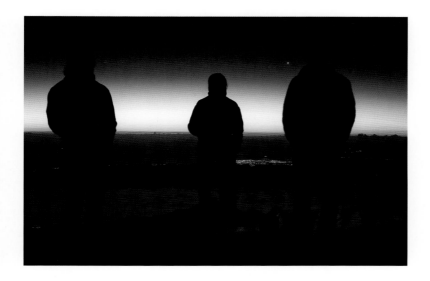

A Stargazer's Delight

Maunakea's natural beauty, dazzling night sky, and astronomical observatories have attracted millions of visitors over the years, making it one of the most popular tourist destinations in Hawai'i. Here three people watch the sunrise while the planets Mercury and Venus are visible in the sky above, together with the lights of Hilo below.

PHOTO: RICHARD WAINSCOAT

The Star Makers

Adaptive optics systems allow earthbound telescopes to see the night sky as clearly as telescopes in space. This technique—perhaps the most revolutionary in astronomy since the invention of photography—works by monitoring a bright star to see how its light is distorted by passage through Earth's turbulent atmosphere and then using this information to remove those distortions, yielding images as sharp as if there were no interfering atmosphere.

But what happens when an astronomer's target of interest has no bright star nearby? In that case, powerful lasers can be used to create artificial stars wherever they're needed. High overhead, at an altitude of about 50 miles (80 kilometers), there's a thin layer of air that's rich in sodium atoms, the ashes of tiny meteorites that burn up as they enter our atmosphere. Lasers with just the right energy can make these sodium atoms glow, creating a bright artificial star in the sky near the object to be observed. As the artificial star's light returns to Earth, it can be monitored by a telescope's adaptive optic system just as if it were a real star.

PHOTO: MICHAEL WEST

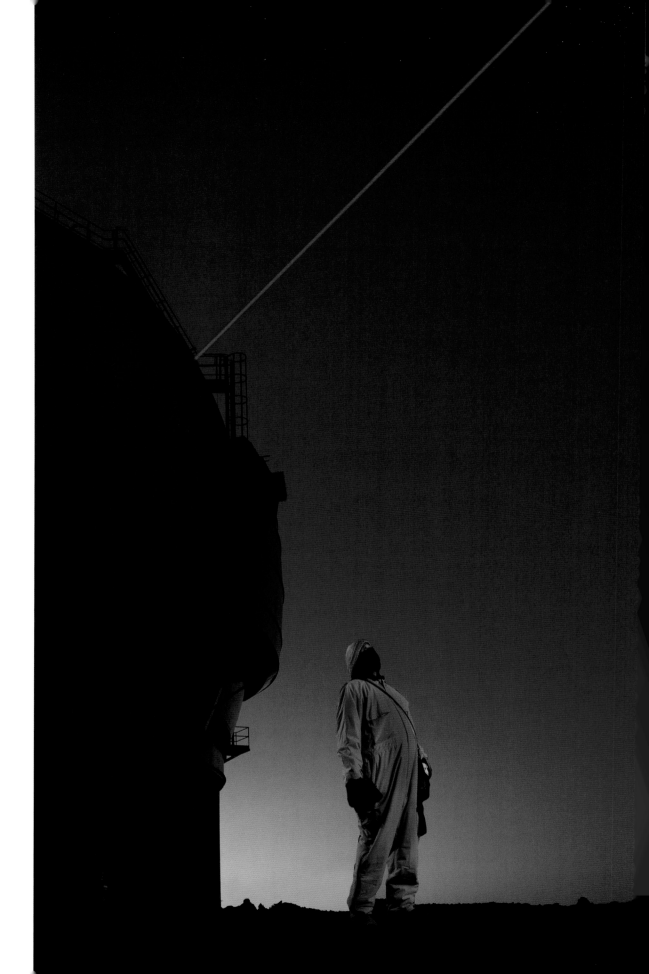

Blue-Sky Laser Guy

A laser fired into the night sky from Keck Observatory creates an artificial star high overhead that's used to monitor and correct distortions that occur whenever light passes through our planet's thick atmosphere. The Gemini and Subaru telescopes use similar laser guide star systems to improve the quality of images.

Although these lasers aren't powerful enough to damage airplanes, care must be taken to ensure that they don't interfere with pilots' vision. In the past, human "spotters" stood outside and scanned the sky from horizon to horizon, shutting down the laser if an approaching aircraft was seen. Automated aircraft spotting systems are now in use. PHOTO: W. M. KECK OBSERVATORY AND RICK PETERSON

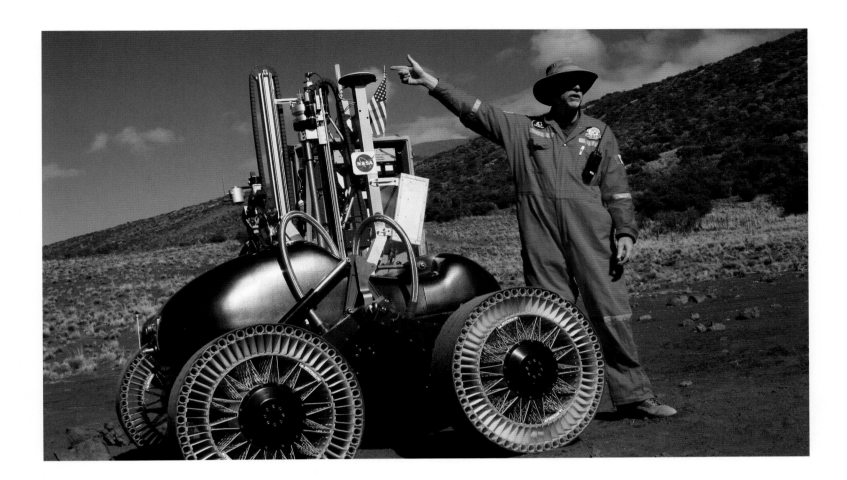

As the last traces of vegetation vanish above an elevation of about 11,000 feet (3,350 meters), Maunakea's barren landscape takes on an otherworldly appearance.

In fact, NASA has long used Hawai'i's volcanic landscape as a testing ground for missions to other worlds. In 1965, Neil Armstrong and fifteen other astronauts spent two weeks training in the island's powdery soil to prepare for the first lunar landings. NASA and its international partners have also come to Maunakea to test technologies for future space missions, including a lunar prospecting robot that might someday allow astronauts to make their own oxygen, water, and fuel by extracting resources from the moon's soil.

Maunakea has similarities with Mars, too. Samples of Martian soil analyzed by the NASA's Curiosity rover in 2012 were found to contain minerals like those found in the volcanic soil on Maunakea's slopes. Such comparisons provide valuable clues about the red planet's past environment, allowing scientist to determine whether Mars might have once been a warmer, wetter world with conditions suitable for life.

Recognizing the island's' unique economic and scientific value as a location for testing emerging aerospace technologies, the State of Hawai'i created the Pacific International Space Center for Exploration Systems in 2007. PISCES, a research and education center headquartered in Hilo, works with NASA, the Canadian Space Agency, and other international partners to conduct environmentally safe field tests of equipment and procedures on the rugged volcanic terrain of Maunakea and nearby Maunaloa.

PHOTO: NASA/PISCES

Same Planet,
Different Worlds

The most beautiful experience we can have is the mysterious—the fundamental emotion which stands at the cradle of true art and true science.

—ALBERT EINSTEIN

POSTCARDS FROM THE UNIVERSE

THE OBSERVATORIES ON MAUNAKEA have been pioneers in astronomical research for more than five decades, blazing new trails into the unknown night after night. They've made countless scientific discoveries during that time, including the first picture of a multiple planet system around another star, the first observations of Kuiper Belt objects in our own solar system, and the first detection of gravitational lensing by galaxies, to name just a few. Without Maunakea's telescopes, our present knowledge of the universe would be considerably poorer.

But the splendors of the night touch the heart as well as the mind, captivating us with their breathtaking beauty. "The sky is the daily bread of the eyes," wrote New England poet and essayist Ralph Waldo Emerson. The telescopes on Maunakea reveal a universe of kaleidoscopic colors and mesmerizing patterns. Like butterfly collectors, astronomers return from their night's work with the captured light of exotic specimens: clouds of interstellar gas sculpted by cosmic winds, galaxies entwined in gravity's grip, and star clusters sparkling like diamonds against the black backdrop of space.

Tonight as the sun sets on Maunakea, another journey into the unknown will begin. Astronomers will make final preparations for the night ahead. Observatory domes will awaken from their daytime slumber and slowly creak open. Telescopes will peer intently into the starry skies until the morning light returns. We live in a golden age of astronomical exploration, with astonishing discoveries being made almost daily. Most of us will never travel into space freed from the shackles of earth's gravity, but thanks to the observatories on Maunakea we can glimpse unimaginably faraway worlds even as we remain bound to our own. The photographs on the following pages are, in a sense, postcards from the universe.

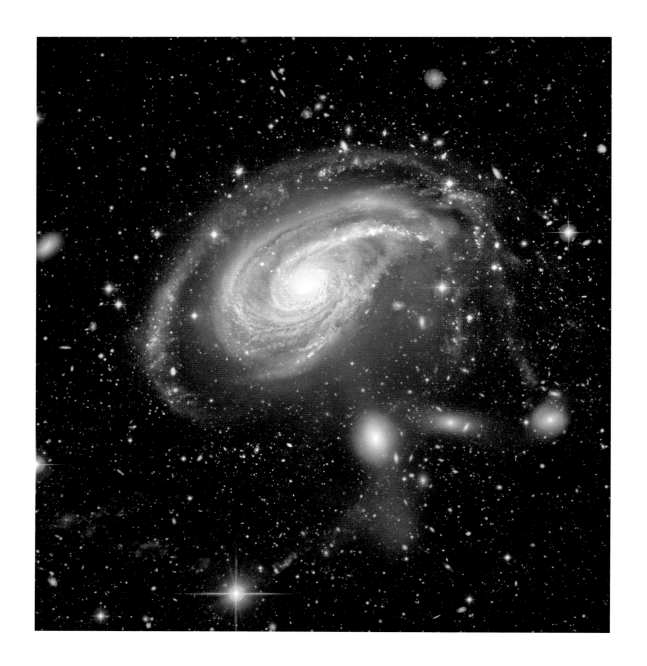

Every Picture Tells a Story

"The universe is made of stories, not of atoms," said poet Muriel Rukeyser.

Every planet, star, and galaxy has a unique story to tell. This image, taken with the 340-megapixel MegaCam camera on the Canada-France-Hawai'i Telescope, captures the light of NGC 772, a galaxy that probably underwent a recent collision with its companion, compact NGC 770, visible near the center of the image. The collision has distorted NGC 772's spiral structure, producing a burst of newborn blue stars and spilling other stars into intergalactic space. NGC 772, also known as Arp 78, is located more than 100 million light-years from our own Milky Way galaxy. IMAGE: CANADA-FRANCE-HAWAI'I TELESCOPE/COELUM, J.-C. CUILLANDRE, AND G. ANSELMI

At the center of the Milky Way, far from the sun, is one of the most hellish environments in the universe. There, two of nature's most extreme objects—a supermassive black hole weighing the equivalent of 4 million suns and an enormous star cluster—conspire to torment gas in their vicinity, blasting it with X-rays, subjecting it to intense gravitational forces, and destroying molecules.

This image, taken with the Submillimeter Array, captures light from three different types of gas molecules that have managed to survive in this region. The dark circle and blue flares represent the black hole and its companion star cluster. Cyanogen (green), formaldehyde (blue), and silicon monoxide (red) molecular gas are each sensitive to different physical conditions, providing information about the diverse processes occurring in different regions. Data from the Very Large Array, a radio telescope in New Mexico, has also been overlaid, showing radio emission from hydrogen gas as it spirals toward the black hole.

Images like this demonstrate the advantages of collecting different types of light, sometimes with different telescopes, as each kind of light provides different information about the universe and its constituents. IMAGE: SERGIO MARTIN RUIZ AND THE SUBMILLIMETER ARRAY

Black Hole Hell

The Art of Darkness

The thirteen telescopes on Maunakea today span a wide range of sizes, designs, and capabilities, and each has contributed in unique ways to broadening our view of the universe.

Instrumentation is often a great equalizer among telescopes. An instrument that analyzes light in novel ways can make a smaller telescope every bit as valuable as a bigger one. A good example is the Canada-France-Hawai'i Telescope (CFHT). Once the largest telescope on Maunakea, CFHT's 11.8-foot (3.6-meter) mirror was eventually surpassed by the light gathering power of the Keck I, Keck II, Subaru, and Gemini North telescopes with their 26- and 33-foot (8- and 10-meter) mirrors. CFHT has remained competitive, however, by developing innovative instruments that provide astronomers with data they can't obtain elsewhere, specializing, for example, in wide field imaging with astronomical cameras capable of capturing the faint light from large portions of the sky.

Resembling a celestial work of art, this cloud of gas and dust known as NGC 2170 reflects the light of nearby stars in beautiful hues. The raw images captured by the telescope were black and white, measuring only the intensity of light coming from different regions of the sky. However, by placing filters in front of the camera and allowing only select colors of light to pass, it's possible to combine exposures into a composite color image. IMAGE: CANADA-FRANCE-HAWAI'I TELESCOPE/COELUM, J.-C. CUILLANDRE, AND G. ANSELMI

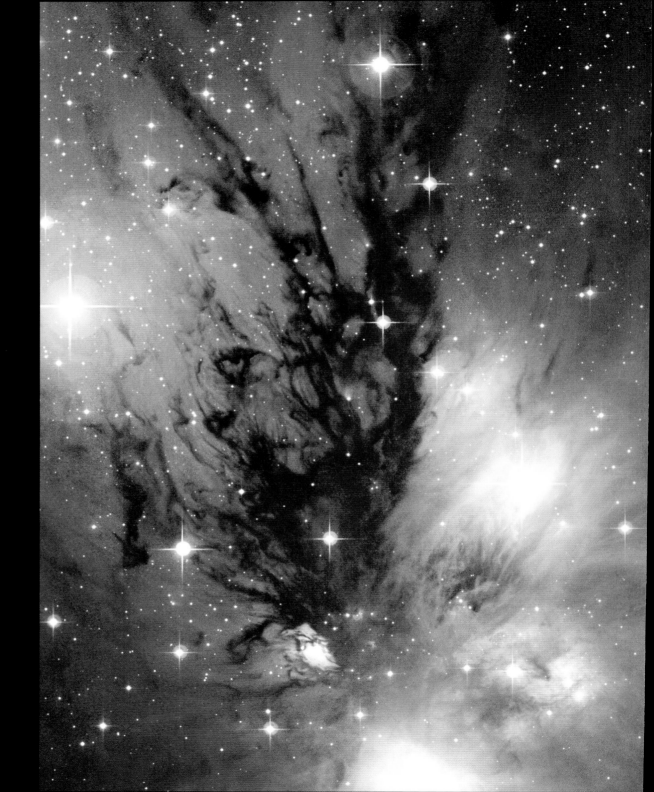

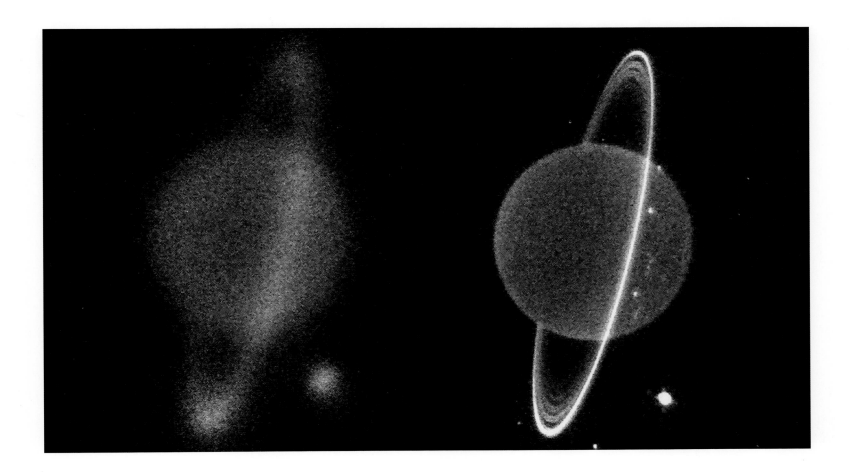

Our planet's atmosphere isn't kind to starlight. Pockets of denser and lighter air distort light rays before they reach telescopes on the ground, blurring images of planets, stars, and galaxies. One of the reasons astronomers come to Maunakea is to get above much of the atmosphere and to take advantage of the calm air over the mountain, which allows sharper images to be obtained here than from most other places on Earth.

Technology makes it possible to further improve on Maunakea's natural assets. A revolutionary development has been the use of adaptive optics to eliminate much of the blurriness caused by our planet's atmosphere. Adaptive optics systems monitor and remove atmospheric distortions as they occur by redirecting light collected by the telescope to a small computer-controlled mirror whose shape changes many times per second to counteract the distortions. The result is images of stunning clarity, as sharp as those obtainable with the Hubble Space Telescope.

These infrared images, taken with the Keck II telescope in 2004, show the planet Uranus with its rings and its moon Miranda as seen in infrared light under normal atmospheric conditions (left) and then using adaptive optics to remove atmospheric blurring (right). IMAGE: W. M. KECK OBSERVATORY, HEIDI B. HAMMEL, AND IMKE DE PATER

A Sharper View

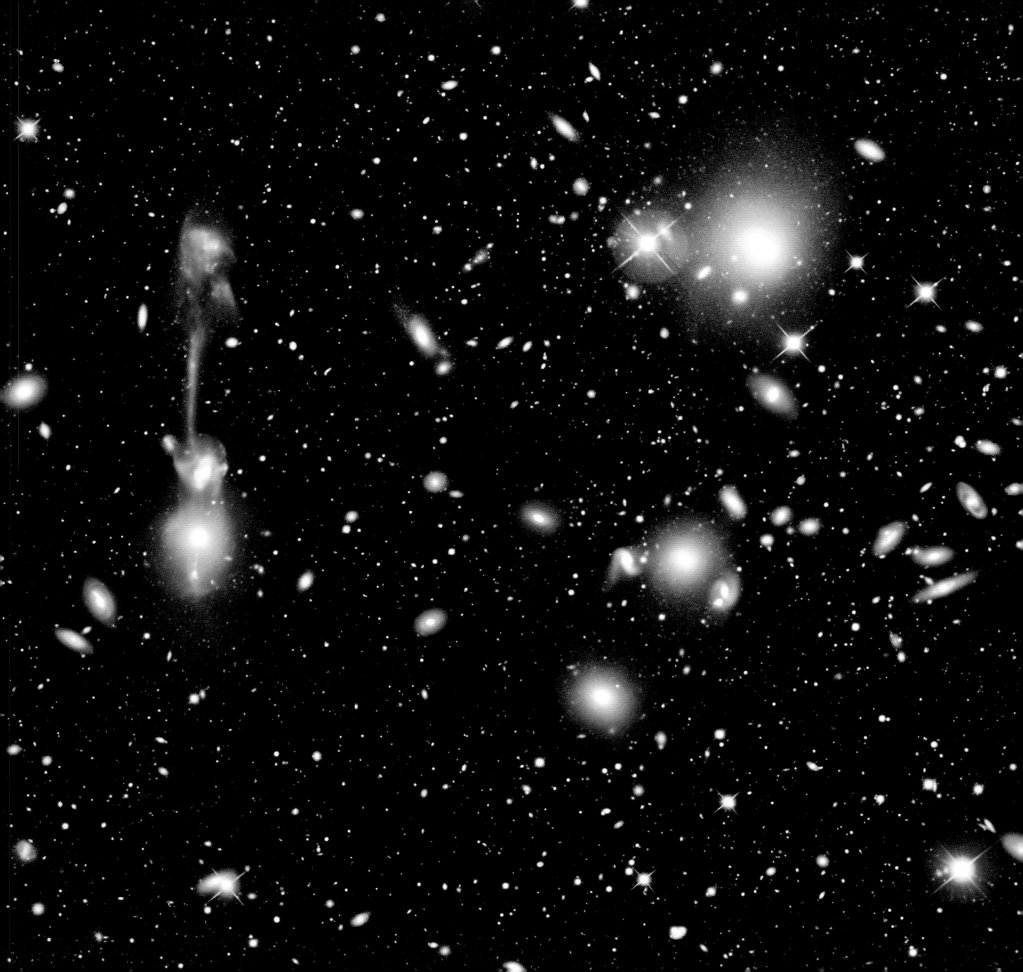

Gregarious Galaxies

Hermits are rare among galaxies, as gravity shepherds them together to form systems ranging from small groups to immense metropolises with thousands of galaxy inhabitants—the cosmic equivalents of New York or Tokyo. This image, captured by the Canada-France-Hawai'i Telescope in 2005, shows a cluster of several thousand galaxies known as Abell 1185 that lies about 400 million light-years away. A dramatic collision between two member galaxies can be seen on the left side of the image. IMAGE: CANADA-FRANCE-HAWAI'I TELESCOPE/COELUM, J.-C. CUILLANDRE, AND G. ANSELMI

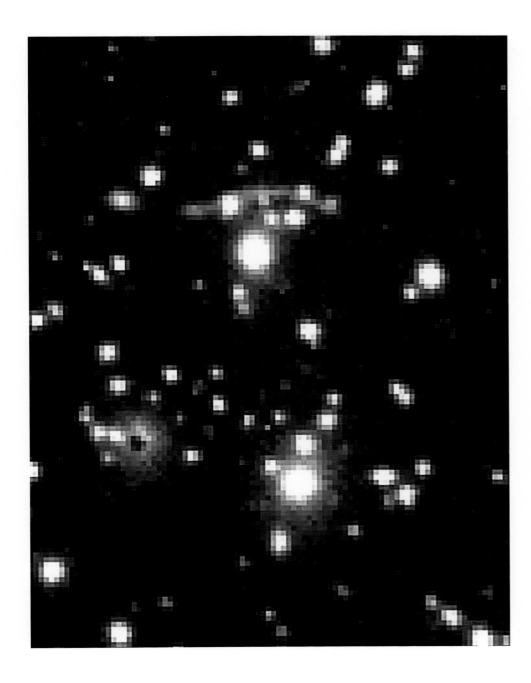

A Maunakea First

There have been many astronomical firsts on Maunakea. One of the most momentous was the first observation of gravitational lensing by a group of galaxies, a prediction of Albert Einstein's General Theory of Relativity.

According to Einstein, massive objects deflect the motion of other objects—including light—that pass nearby. This leads to a phenomenon known as gravitational lensing, in which images of distant objects become distorted as their light travels near other objects en route to Earth. Einstein himself thought that this effect would be too small to be seen. But he was wrong.

In 1985, astronomers using the Canada-France-Hawai'i Telescope snapped the first image of a giant gravitational arc, the lensed image of a distant galaxy whose light has been bent as it passes through a group of galaxies known as Abell 370. The group's strong gravity acts like an enormous funhouse mirror, stretching and distorting the more distant galaxy's image to produce an optical illusion on the grandest scale. This historic image was obtained using CFHT's first digital camera, which had only 320 x 512 pixels, far fewer than today's consumer digital cameras.

Gravitational lensing has become one of astronomy's most powerful research tools, revealing the presence and distribution of enormous quantities of invisible matter in the universe as well as insights into some of the most distant objects ever seen. IMAGE: CANADA-FRANCE-HAWAI'I TELESCOPE

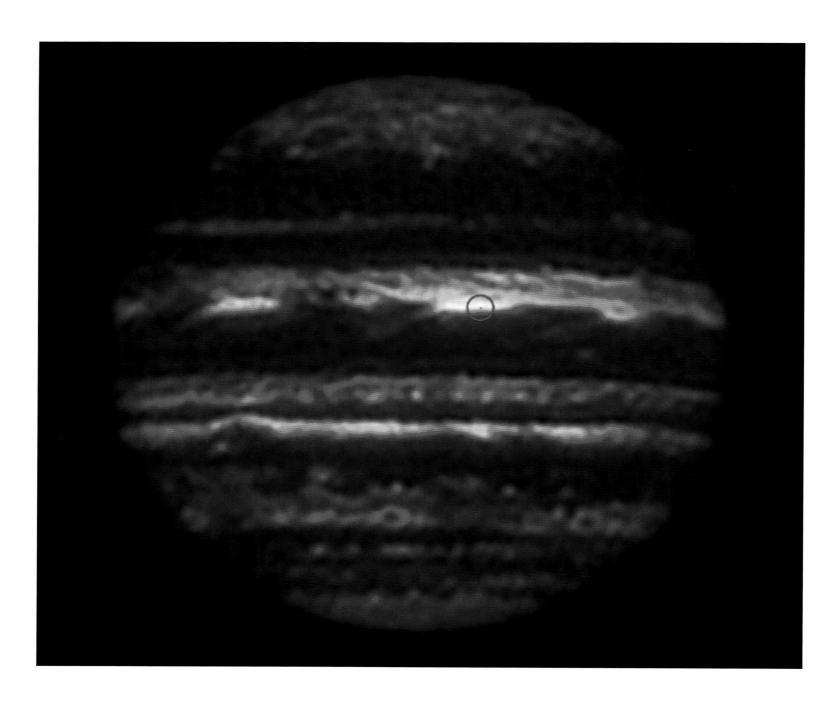

Support for Space Missions

The NASA Infrared Telescope Facility (IRTF) on Maunakea was built to provide ground-based observations in support of NASA space missions. This infrared image of Jupiter, taken by the IRTF in 1995, shows where heat is emerging from the lower atmosphere of Jupiter. The blue circle indicates the location where a probe released by the Galileo spacecraft plunged into the giant planet's atmosphere to learn more about conditions there. The IRTF observations showed that the probe entered a relatively cloud-free region, important information for interpreting the data sent back. IMAGE: IRTF AND GLENN ORTON, JET PROPULSION LABORATORY (JPL)

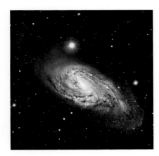

Cosmic Storm

"The whole universe appears as an infinite storm of beauty," wrote the nineteenth-century naturalist John Muir.

M66, a spiral galaxy about 36 million light-years away, probably looks quite similar to our own Milky Way galaxy. This image, taken with the Canada-France-Hawai'i Telescope, reveals its delicate swirling structure that resembles a hurricane in space. Dark streaks visible in the galaxy are regions of dust, debris from generations of stars long gone that will provide the raw material for future generations of stars, planets, and maybe life.

The observatories on Maunakea have been at the forefront in the study of galaxies, providing many new insights into their origins, evolution, and properties. IMAGE: CANADA-FRANCE-HAWAI'I TELESCOPE/COELUM, J.-C. CUILLANDRE, AND G. ANSELMI

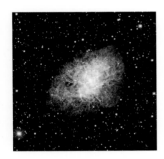

Death of a Star

It was a heart attack waiting to happen. A massive star that probably weighed ten times more than our sun died suddenly and spectacularly. The Crab Nebula, seen in this image from the Canada-France-Hawai'i Telescope, is the aftermath of the star's explosive death that was first observed in 1054.

Heavyweight stars like this live fast and die young, burning with such intensity that they quickly burn out, while smaller stars like our sun smolder for much longer. To put it in perspective, if we imagine that the sun had a typical human lifespan of seventy-five years, then the entire life of the star that produced the Crab Nebula would be measured in weeks, not years.

The brief but intense lives and explosive deaths of stars like this one enrich the universe in ways our more placid sun never can, creating the calcium in our bones, the aluminum that coats telescope mirrors, and the seeds of future worlds. IMAGE: CANADA-FRANCE-HAWAI'I TELESCOPE/COELUM, J.-C. CUILLANDRE, AND G. ANSELMI

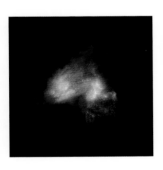

A Supernova Factory

This object, known as Arp 299, is actually the ongoing merger of two galaxies. Observations made with the Very Long Baseline Array (VLBA) in 2002 and 2003 revealed an abundance of supernovae—the explosive deaths of massive stars—hidden deep inside it. The birth of these heavyweight stars was triggered millions of years earlier when the two galaxies began to collide. Because of obscuring dust within Arp 299 and its distance from Earth, only a radio telescope with the VLBA's penetrating ability to see fine detail could reveal these hidden supernovae.

This image combines radio light from the VLBA and Green Bank Telescope in West Virginia together with visible light collected by the Hubble Space Telescope. IMAGE: B. SAXTON FROM DATA PROVIDED BY J. ULVESTAD, S. NEFF, AND S. TENG; NRAO/AUI/NSF; AND NASA/ESA HST

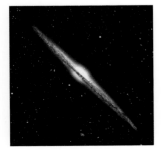

Island in the Sky

Galaxies appear like islands of light shining in the dark ocean of space. This galaxy, known as NGC 4565, is located about 40 million light-years away, which means we see it as it looked 40 million years ago. Discovered by astronomer William Herschel in 1785, it's a spiral galaxy seen edge-on. Our Milky Way galaxy would look quite similar, though slightly smaller, if seen from the same perspective. IMAGE: CANADA-FRANCE-HAWAI'I TELESCOPE/COELUM, J.-C. CUILLANDRE, AND G. ANSELMI

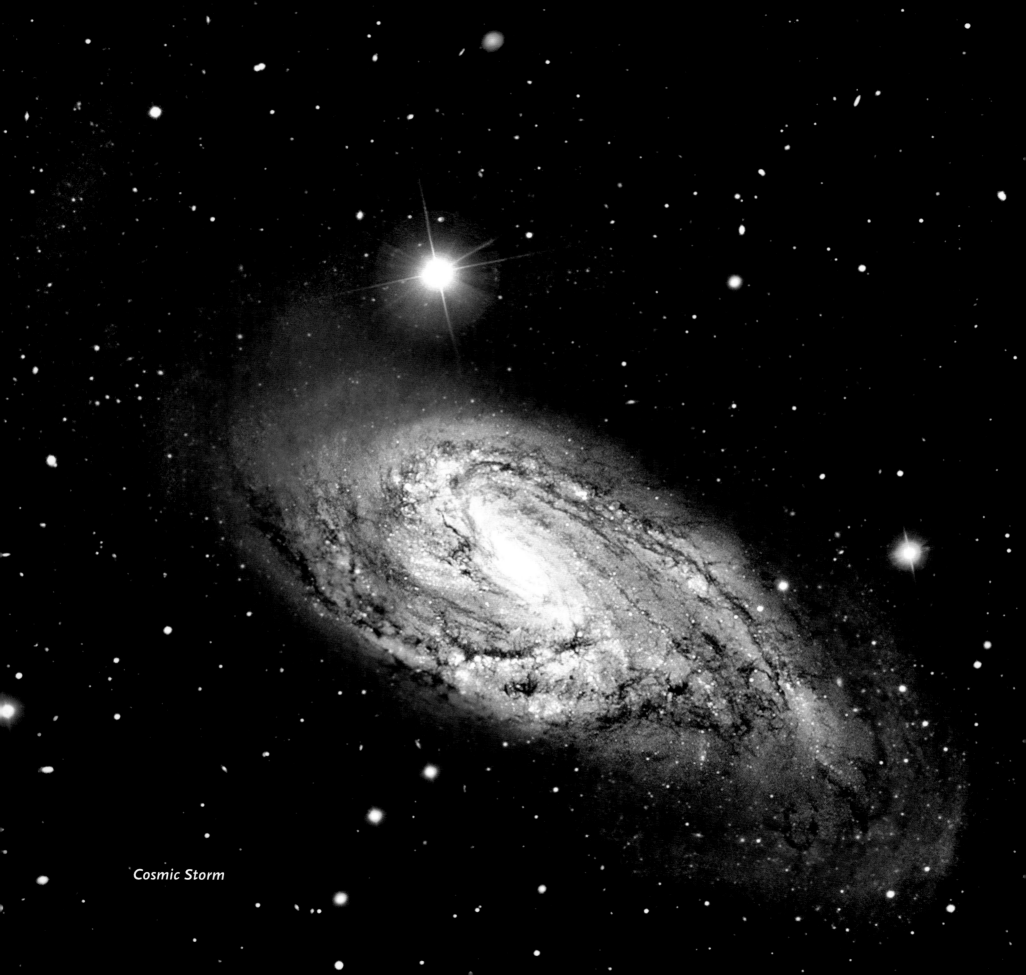

Cosmic Storm

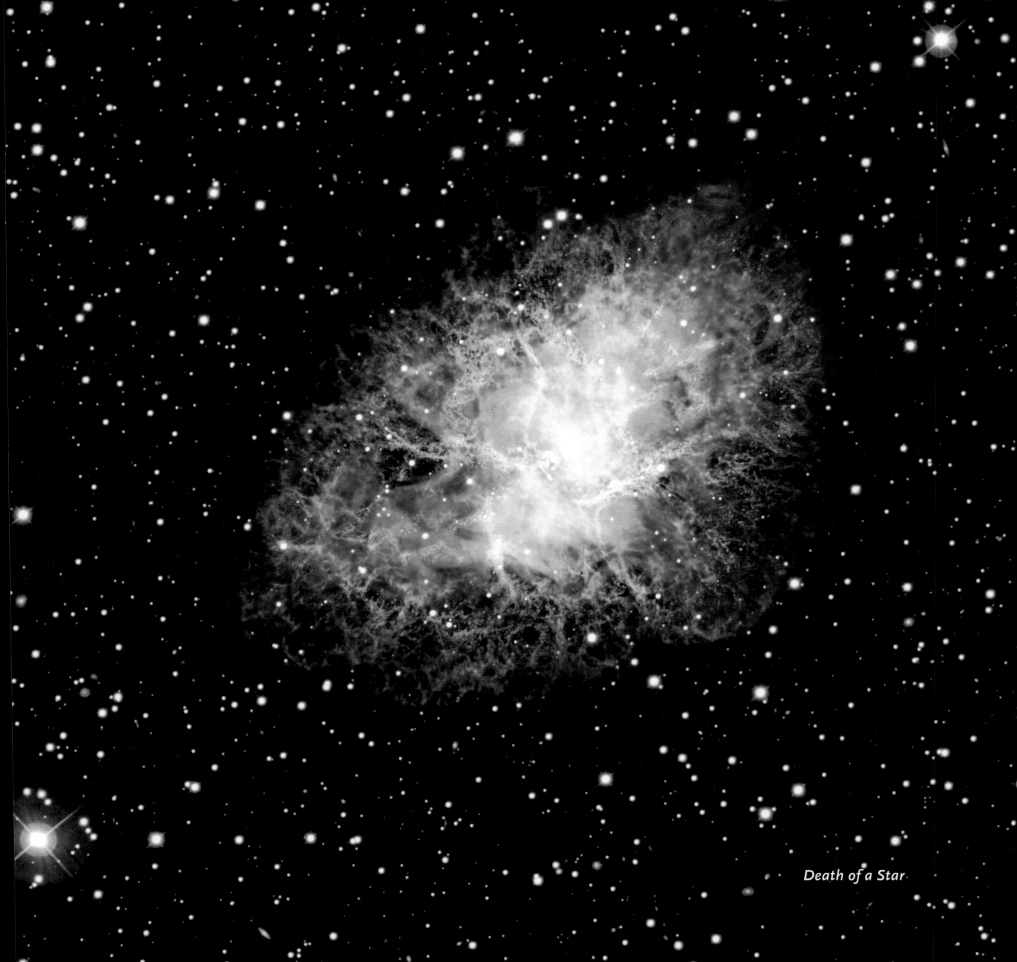

Death of a Star

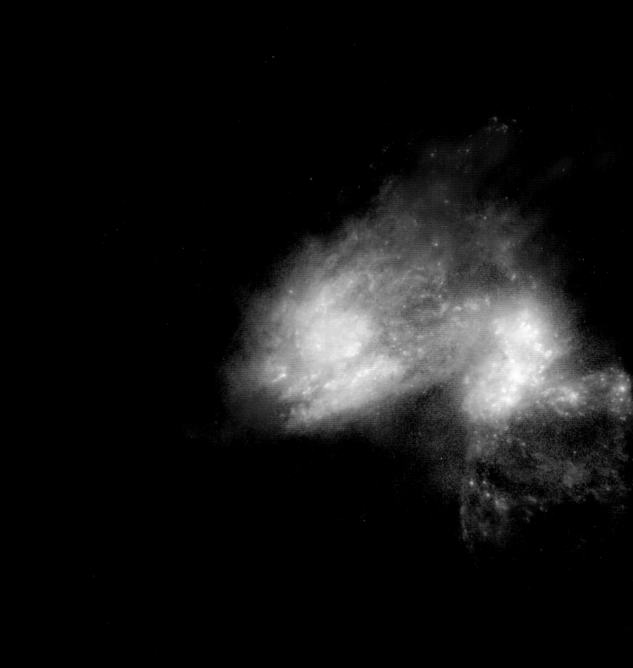

A Supernova Factory

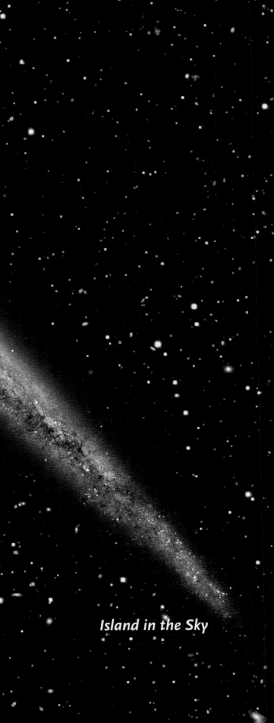

Island in the Sky

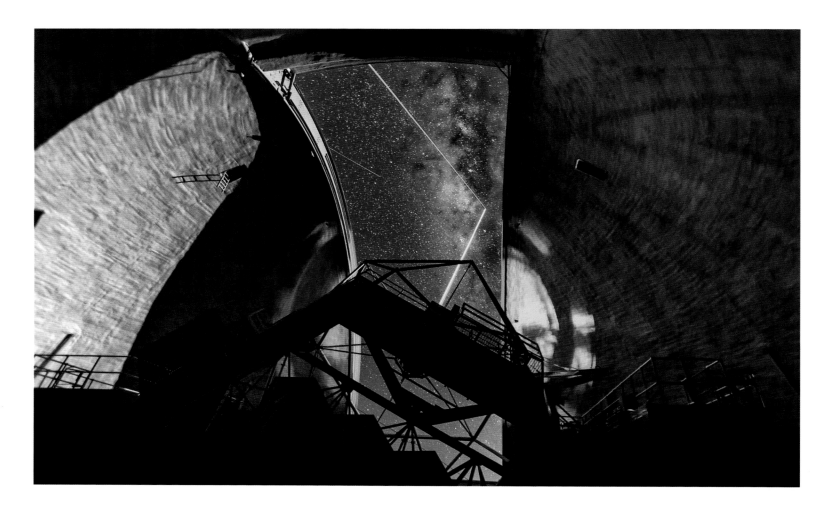

The Beast in the Heart of Our Galaxy

An enormous black hole lurks within the heart of the Milky Way. Astronomers can't see it, because it emits no light, but its gravitational grip on stars that stray too close is obvious.

For two decades, the Keck Observatory telescopes have tracked the orbits of stars whose paths are deflected by the gravitational pull of a massive but invisible object near the center of our galaxy. One star in particular has been observed whizzing around this point in space every 11.5 years. Astronomers use this information to weigh the invisible object, calculating how heavy it must be for its gravity to make these stars move as fast as they do. This dark object turns out to weigh as much as 4 million suns—astronomers call it a supermassive black hole.

Our galaxy isn't the only one with a supermassive black hole. Keck Observatory, Gemini, Subaru and other telescopes on Maunakea have found them in the centers of other galaxies too, with bigger galaxies having bigger black holes. How these behemoths—which far outweigh even the largest known star—came into existence remains a mystery.

This image shows the laser guide star systems of the Keck I and Keck II telescopes in action while astronomers observe the center of the Milky Way. The lasers create artificial stars high overhead that are used to monitor and reduce blurriness that occurs when light passes through our planet's atmosphere, allowing the sharpest and most detailed images of the region around our galaxy's enormous black hole to be obtained. PHOTO: W. M. KECK OBSERVATORY AND ETHAN TWEEDIE

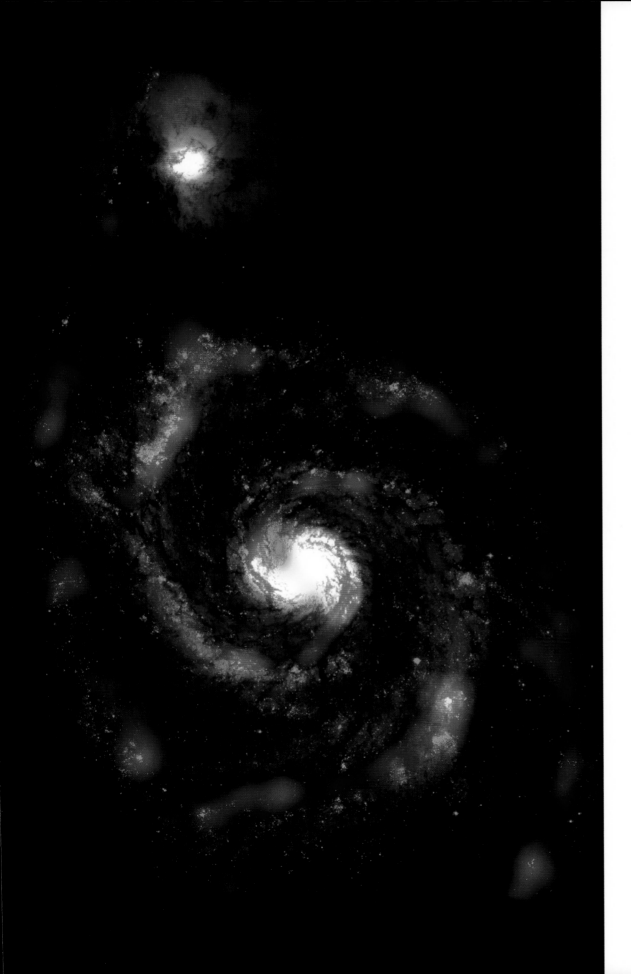

A Whirlpool of Colors

"Step by step astronomy wrests fresh secrets from the starry abysses." These words, written by the nineteenth-century English poet and journalist Edwin Arnold, are as true today as they were then.

In astronomy, like all fields of science, new discoveries build on previous ones. This composite image of the Whirlpool Galaxy combines images taken with two different telescopes seven years apart. The green image was captured by the Hubble Space Telescope in 2005 and shows visible light emitted by the galaxy. Submillimeter light detected by the James Clerk Maxwell Telescope in 2012 is shown in red and blue, with each color corresponding to light of different amounts of energy. This submillimeter light is emitted by cold dust, the raw material from which new generations of stars form. IMAGE: JOINT ASTRONOMY CENTRE, UNIVERSITY OF BRITISH COLUMBIA, AND NASA/HST (STScI)

A Litter of Stars

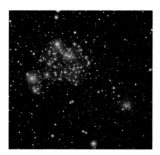

Stars, like puppies, are born in litters. We humans usually give birth to one offspring at a time. For us, the birth of twins is rare and triplets even rarer. But for stars, multiple births are the norm. In fact, it's unusual for a star to be born without siblings—lots and lots of siblings.

The birth of stars takes place in giant clouds of gas and dust that are slowly squeezed by gravity until they condense into stars. Depending on its size, a single cloud can spawn dozens to millions of stars. These litters usually disperse over time as siblings drift apart, tugged this way and that by the gravitational pull of other passing stars, star clusters, and gas clouds. Most stars manage to hang onto a sibling or two, however, becoming lifelong companions in an endless gravitational embrace.

The sun was probably born in such a litter nearly 5 billion years ago. Circumstantial evidence suggests that our star might have had a thousand or more siblings. But where they are today is anybody's guess. Like a litter of puppies adopted by different families, the sun's siblings have spread to parts unknown throughout our galaxy.

This image, captured in a fifteen-minute exposure by the MegaCam camera of the Canada-France-Hawai'i Telescope (CFHT) in 2003, shows two star clusters of very different ages. The blue stars to the left belong to a cluster known as M35, which is believed to have formed about 100 million years ago. The yellowish clump of stars to the right, known as NGC 2158, is about ten times older and much farther away. Hundreds of thousands of stars are visible in this image.

Although it's no longer the largest telescope on Maunakea, CFHT has carved out a niche by specializing in wide-field imaging, producing exquisite panoramic views of the heavens.

IMAGE: CANADA-FRANCE-HAWAI'I TELESCOPE/COELUM, J.-C. CUILLANDRE, AND G. ANSELMI

A Dying Star

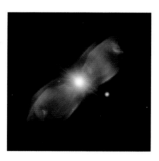

This infrared image of a dying star was taken with the Gemini North telescope in 2005. As the aging star began to run out of fuel, it gently expelled its outer regions into space like a dying gasp. Similar objects are seen throughout space. Called planetary nebulae for historical reasons—although they have nothing to do with planets—these dying stars create some of the most beautiful sights in the night sky. Our sun, which is a middle-aged star, will probably meet a similar fate in 4 to 5 billion years. IMAGE: GEMINI OBSERVATORY/AURA

The Sounds of the Night

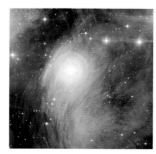

At night the sky over Maunakea seems to come alive with stars, painting the darkness with a rainbow of colors.

Yet it's not only the view of the heavens that captivates, it's the silence, too—a sort of cosmic hush. We've become so accustomed to noise in our lives that complete silence can be jarring. Gazing into the starry skies in the stillness of Maunakea, there's an almost overwhelming sense of connection, a feeling that we're part of something much bigger and more wondrous than we can even imagine. In the quiet of the night on Maunakea, it's hard not to feel as if we're witnessing that moment when an artist contemplates his or her creation in silence and knows that everything is just as it should be. IMAGE: CANADA-FRANCE-HAWAI'I TELESCOPE/COELUM, J.-C. CUILLANDRE, AND G. ANSELMI

A Litter of Stars

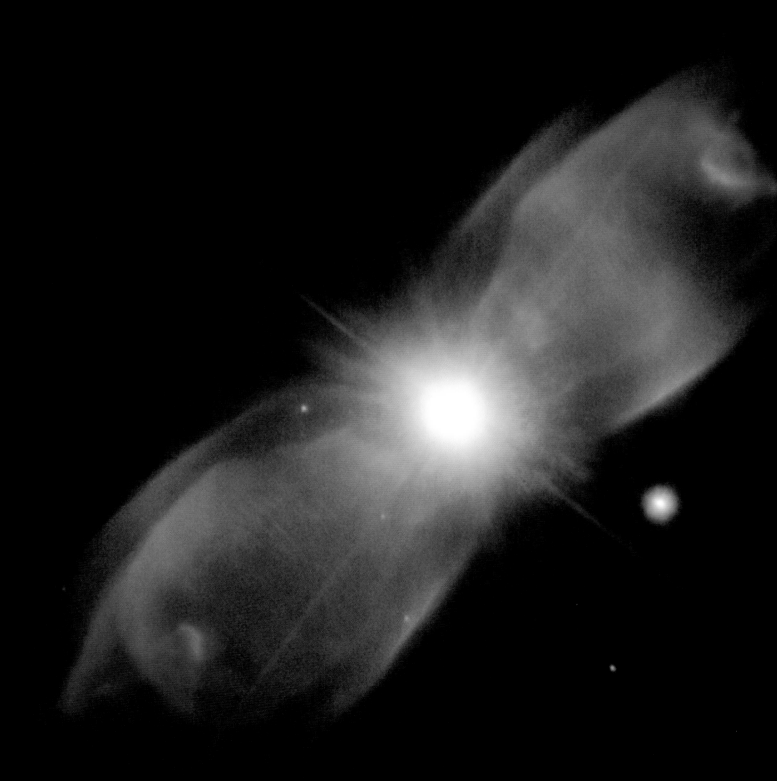

A Dying Star

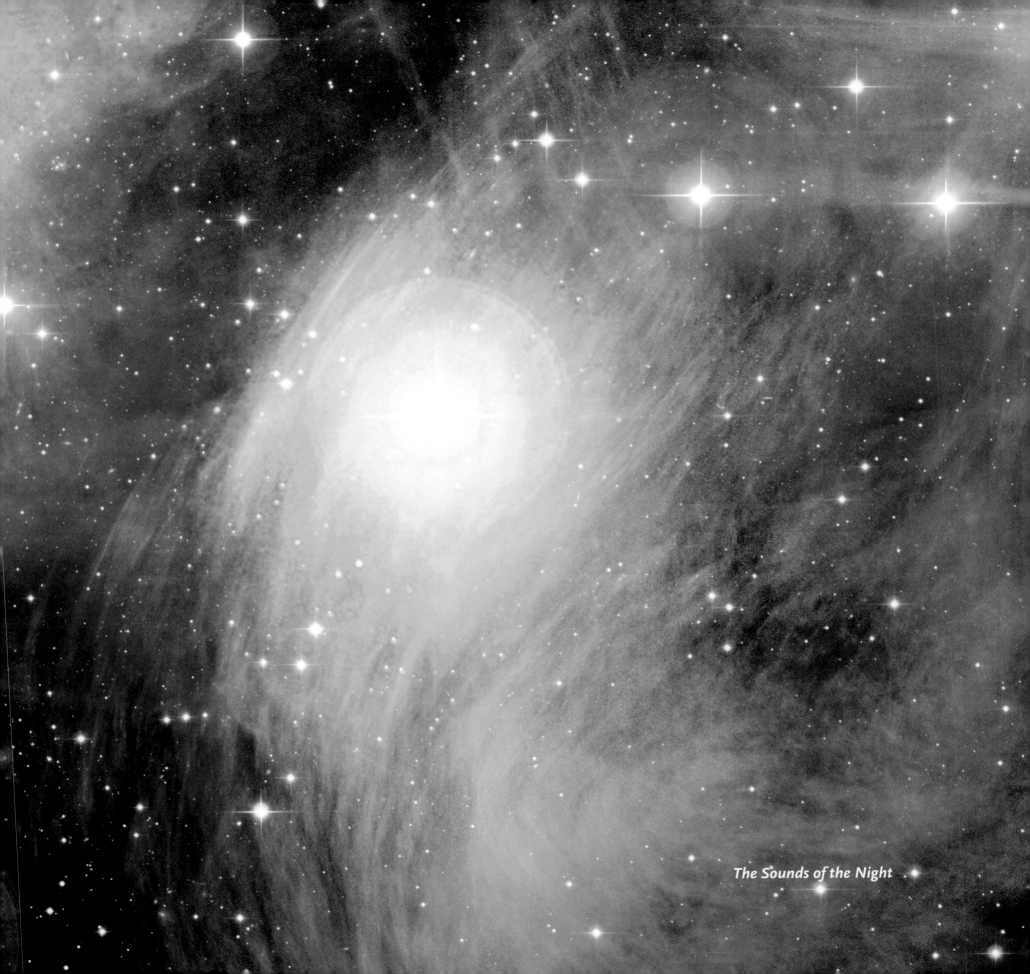

The Sounds of the Night

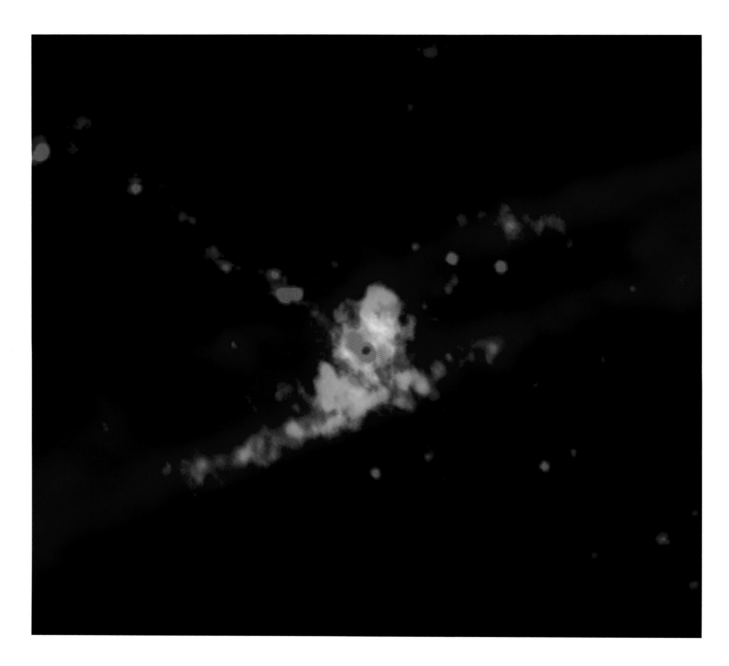

Feeding the Monster

Like many galaxies, Centaurus A (also known as NGC 5128) has an enormous black hole at its center that weighs millions of times more than our sun. As the black hole's gravity pulls in surrounding gas, narrow jets of material are ejected at high speeds, for reasons not yet fully understood. The Submillimeter Array (SMA) peered into the heart of Centaurus A to capture this remarkable photo of the black hole's food supply. Much of the gas is in a rotating disk around the black hole (green), but a more elongated distribution spanning tens of thousands of light-years is also seen (blue). X-ray light coming from the black hole's jet, captured by the Chandra space telescope, is shown in red. IMAGE: DANIEL ESPADA AND THE SUBMILLIMETER ARRAY

160

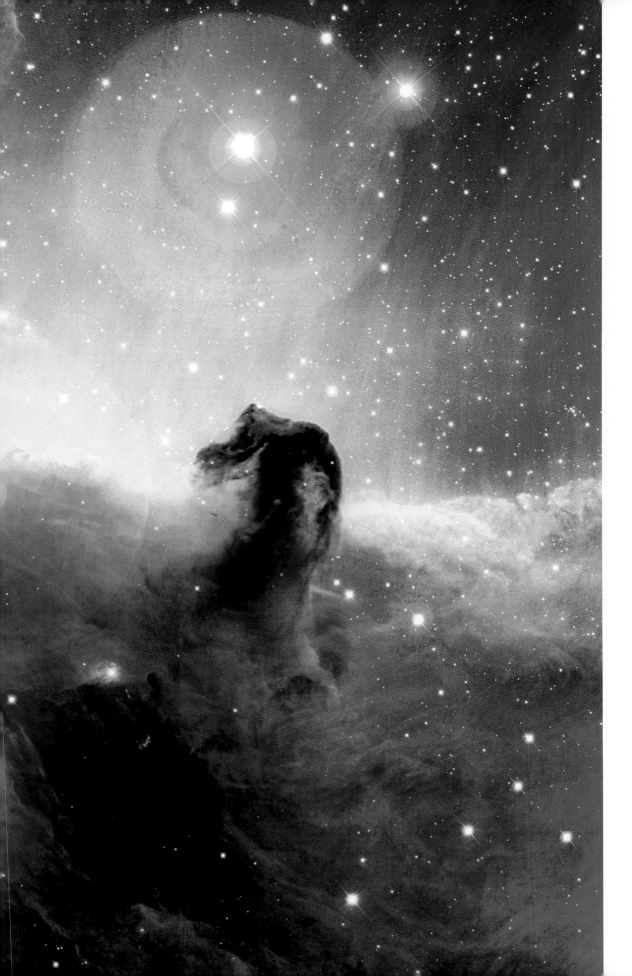

A Cosmic Rorschach Test

The beautiful Horsehead Nebula, named for its obvious equine appearance, is a stunning example of nature imitating art. This cold, dark cloud of gas and dust has been sculpted over eons by light from nearby stars and gusty stellar winds. Located about 1,500 light-years from Earth, the Horsehead Nebula is silhouetted against the backdrop of a glowing region of gas known as IC 434.

The Canada-France-Hawai'i Telescope (CFHT) snapped this spectacular image of the Horsehead Nebula in all its artistic splendor in 2003. IMAGE: CANADA-FRANCE-HAWAI'I TELESCOPE/COELUM, J.-C. CUILLANDRE, AND G. ANSELMI

The Heart of the Milky Way

This composite image combines light from three different telescopes—the Caltech Submillimeter Observatory (CSO, orange), the Very Large Array (VLA, purple), and NASA's Spitzer Space Telescope (cyan)—to give a remarkably detailed view of the center of our galaxy.

The brightest region on the left half of the image is a stellar nursery called the Sagittarius B2 complex, perhaps the most active region of star formation in the Milky Way today. The brightest region in the right half of the image contains the Sagittarius A complex, where a supermassive black hole—several million times more massive than our sun—lies hidden behind the glow of gas and dust.

Images like this demonstrate the power of collecting different kinds of light, as each provides different insights into the night sky. Radio waves collected by the VLA trace regions where gas is being illuminated by hot, young stars. Submillimeter light detected by CSO reveals the distribution of cold gas and dust that will form new stars in the next few million years. And the infrared light captured by Spitzer is primarily from stars and organic molecules. Together, these observations provide one of the most detailed looks into the heart of the Milky Way that has ever been obtained. IMAGE: ADAM GINSBURGH/CSO/NRAO/AUI/NASA/JPL-CALTECH

Cosmic Cartographers

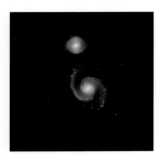

One of the biggest changes in the way science is done on Maunakea over the past few decades has been the investment of large amounts of telescope time in surveys of the sky that can be shared by astronomers for a wealth of different scientific goals.

A good example is the Canada-France-Hawai'i Telescope Legacy Survey (CFHTLS) undertaken from 2003 to 2009, which devoted 450 nights to imaging large portions of the sky with CFHT's 340-million-pixel MegaCam camera. More than 22,000 individual exposures were taken, and the resulting data have been used to study topics ranging from our solar system to the distant universe.

Another example is the United Kingdom Infrared Telescope (UKIRT) Infrared Deep Sky Survey (UKIDSS), which began in 2005 with the goal of using UKIRT's wide-field infrared camera to map a large fraction of the northern sky. The very first UKIDSS survey image is shown here, a patch of sky containing the beautiful Whirlpool Galaxy. Hundreds of scientific papers have been published based on the UKIDSS and CFHTLS surveys. IMAGE: JOINT ASTRONOMY CENTRE/UKIDSS

A Close Encounter

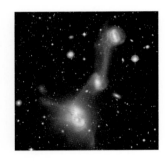

Seen from the perspective of billions of years, galaxies are as transient as clouds on a summer day. Images taken with telescopes on Maunakea reveal large galaxies devouring smaller ones in acts of cosmic cannibalism. Other galaxies are seen being ripped apart by sudden gusts of gravity from passing neighbors, scattering their fragmented remains throughout space in beautiful arcs of starlight. The NGC 4410 galaxy group, seen here, consists of about a dozen members, several of which are interacting strongly. IMAGE: CANADA-FRANCE-HAWAI'I TELESCOPE/COELUM, J.-C. CUILLANDRE, AND G. ANSELMI

Celestial Fireworks

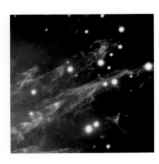

George Lemaitre, considered the father of Big Bang cosmology, wrote, "The evolution of the world can be compared to a display of fireworks that has just ended: some few red wisps, ashes, and smoke. Standing on a well-chilled cinder, we see the slow fading of the suns and we try to recall the vanished brilliance of the origin of the worlds."

The Gemini North telescope captured this spectacular display of celestial fireworks going on some 1,500 light-years away. The image, recorded in infrared light, is remarkably sharp thanks to Gemini's adaptive optics system, which uses a special shape-shifting mirror to cancel out distortions that occur whenever light passes through our planet's turbulent atmosphere.

The blue blobs of glowing material have been ejected from a nearby region filled with energetic newborn stars. These blobs leave colorful wakes as they move at speeds of hundreds of miles per second through surrounding gas.

Gemini observatory has been a pioneer in adaptive optics. Building on experience gained on Maunakea, in 2014 the observatory inaugurated the world's largest and most sophisticated adaptive optics system for use on the Gemini South telescope in Chile. The Gemini Multi-Conjugate Adaptive Optics System (GeMS) uses multiple lasers and shape-shifting mirrors to provide even sharper images over wider areas of sky. IMAGE: GEMINI OBSERVATORY/AURA

Cosmic Cartographers

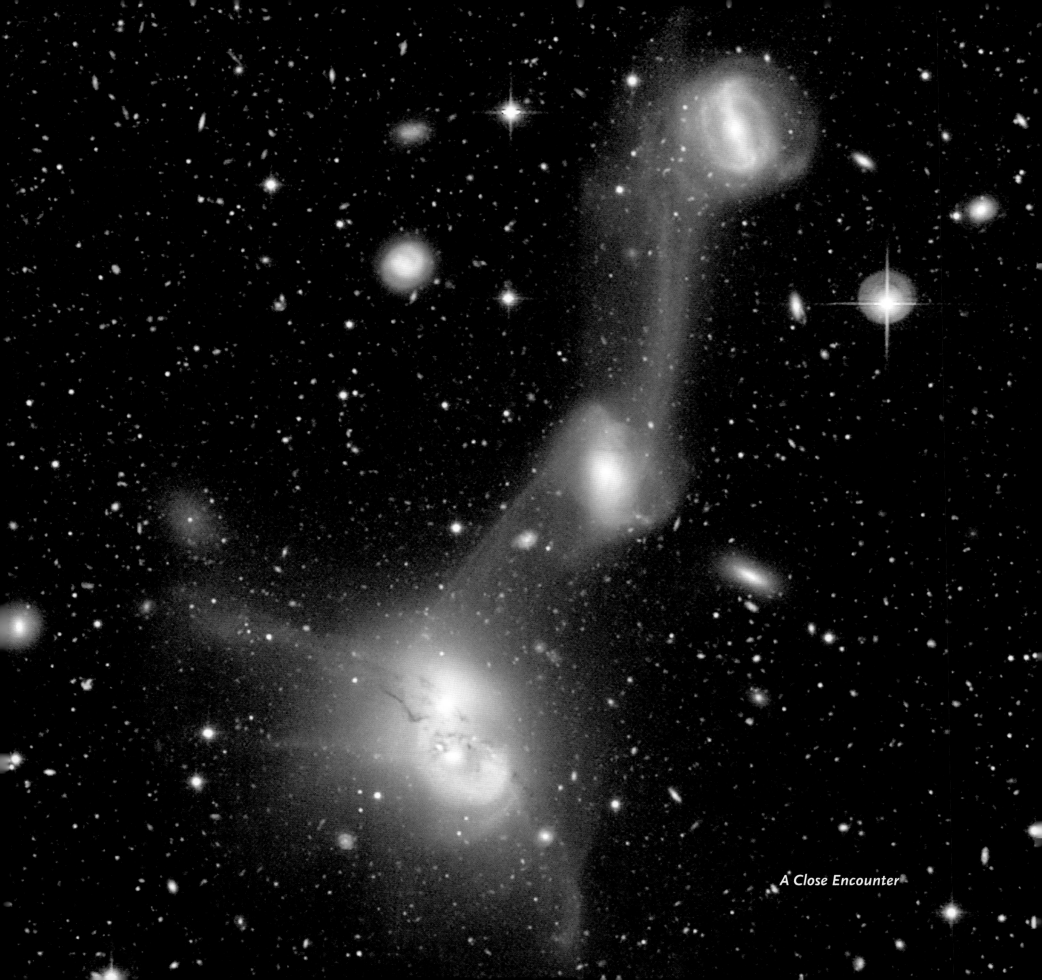

A Close Encounter

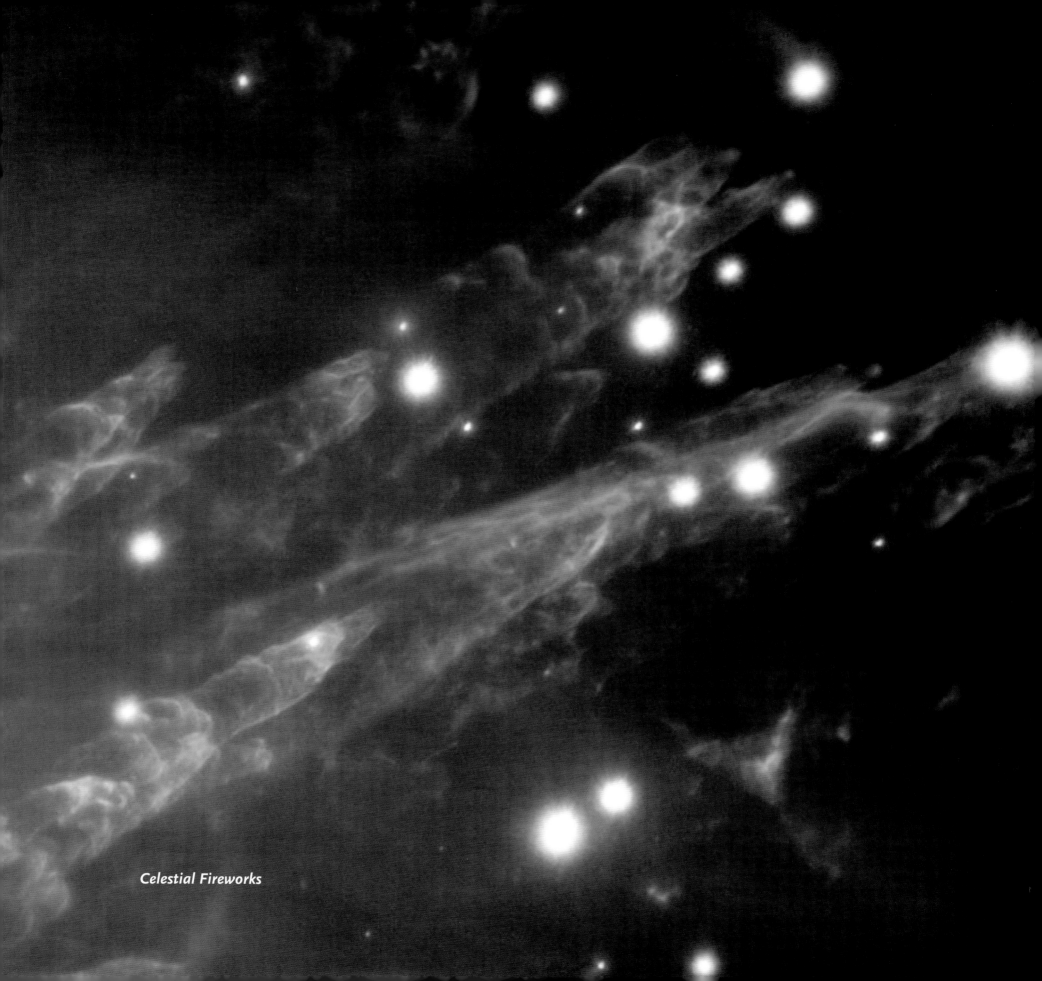
Celestial Fireworks

More Stars than People

There are several hundred billion stars in our galaxy—more than the total number of people who ever lived. Our sun is just one of them, residing in the suburbs of the Milky Way at a distance of about 27,000 light-years from the center.

Unfortunately, the space between stars contains tiny dust grains that block many kinds of light, including the visible light our eyes see, which obscures our view of the galaxy's center. Infrared light and radio waves, however, are not impeded by this dust and hence provide relatively unobstructed views of our galaxy.

This image, taken with the United Kingdom Infrared Telescope (UKIRT), captures the infrared light of myriad stars in a small patch of sky towards the center of the Milky Way. The dark streaks are regions where interstellar dust is so thick that even infrared light can't penetrate them, creating an opaque silhouette against the bright infrared glow of background stars. IMAGE: JOINT ASTRONOMY CENTRE AND PHIL LUCAS ET AL.

A Watery World

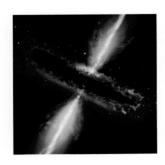

In an ironic twist, astronomers who came to Maunakea because of its dry air discovered one of the wettest places in the universe.

In 2008, astronomers used the Caltech Submillimeter Observatory (CSO) to observe a quasar known as APM 08279+5255 located more than 12 billion light-years away. Quasars are some of the most luminous objects in the universe. Fueled by supermassive black holes that devour anything that strays too close, they release enormous amounts of energy in the process. APM 08279+5255 is home to a black hole that's 20 billion times more massive than the sun and produces as much energy as a thousand trillion stars each second.

CSO's Z-Spec instrument revealed that APM 08279+5255 is surrounded by the largest and most distant reservoir of water ever seen in the universe. The water, equivalent to 140 trillion times all the water in Earth's oceans, is in the form of water vapor because of the intense light coming from the quasar.

Submillimeter telescopes such as CSO need dry skies overhead because water vapor in our planet's atmosphere absorbs much of the submillimeter light from space before it reaches telescopes on the ground. Maunakea's peak is above nearly half our planet's atmosphere, and the remaining air overhead is quite dry, making it an ideal site for telescopes like CSO.

This artist's illustration shows what the disk of material that feeds the supermassive black hole in APM 08279+5255 might look like. IMAGE: NASA/ESA

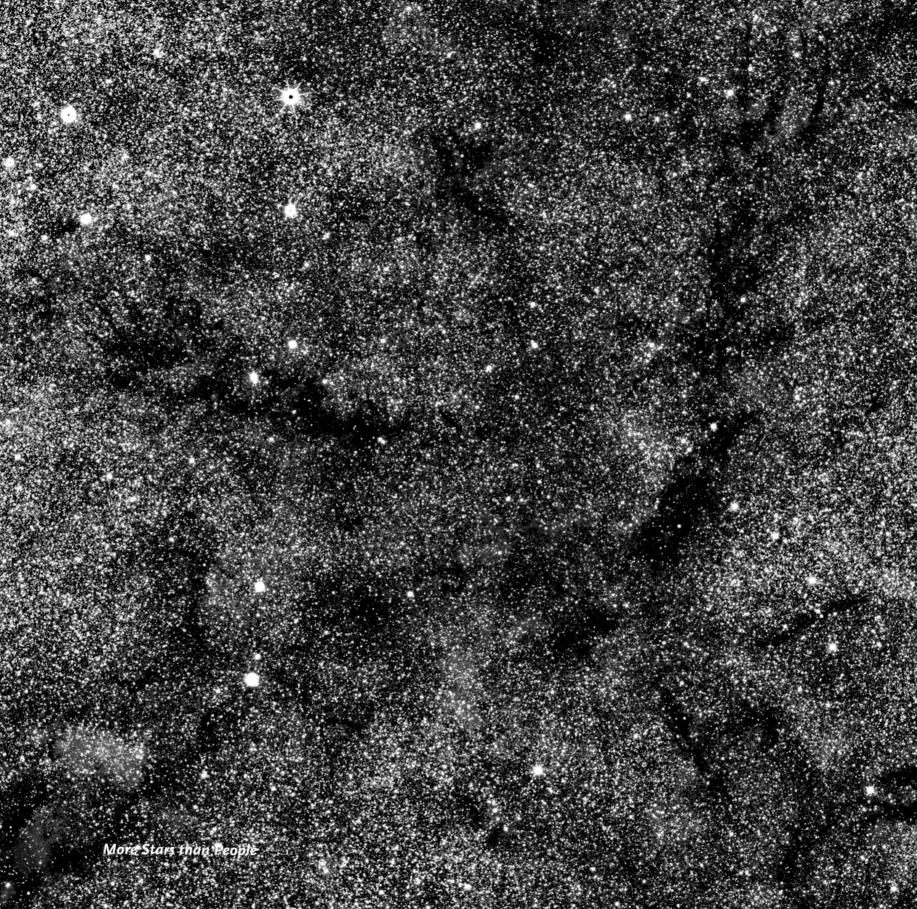
More Stars than People

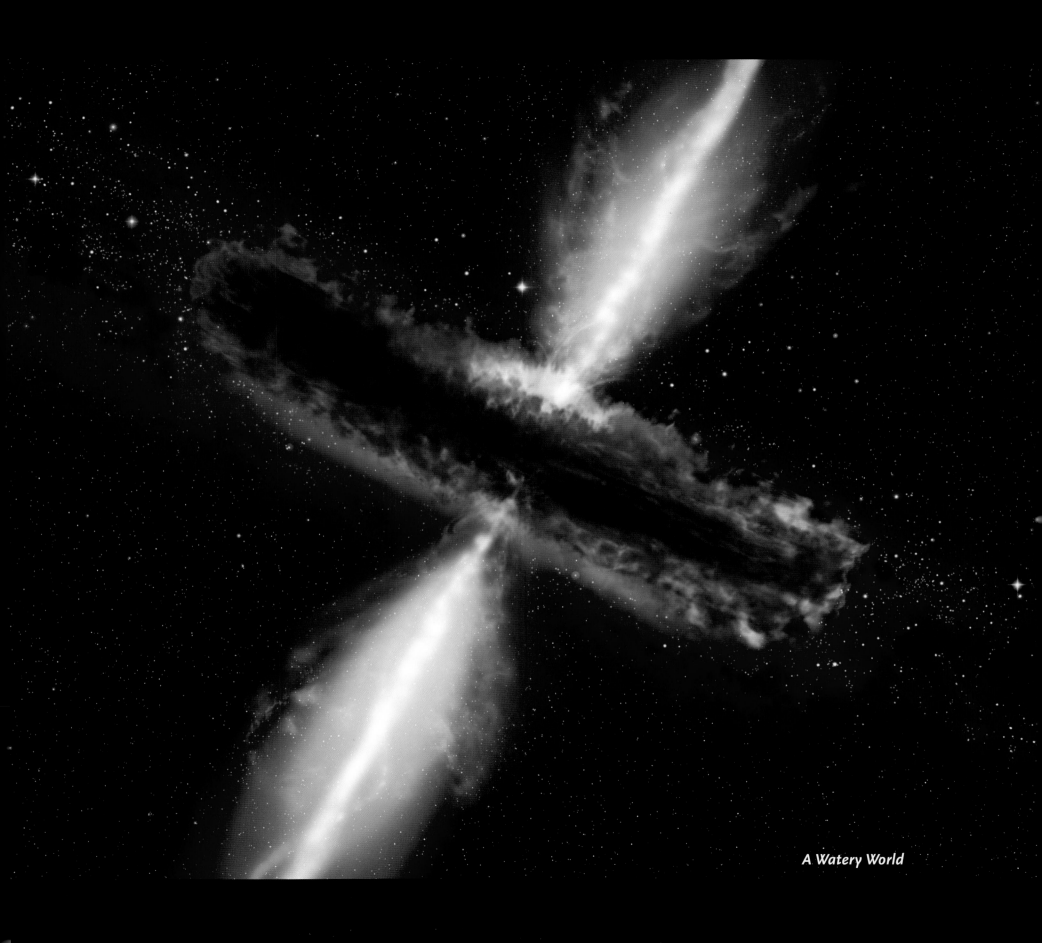

A Watery World

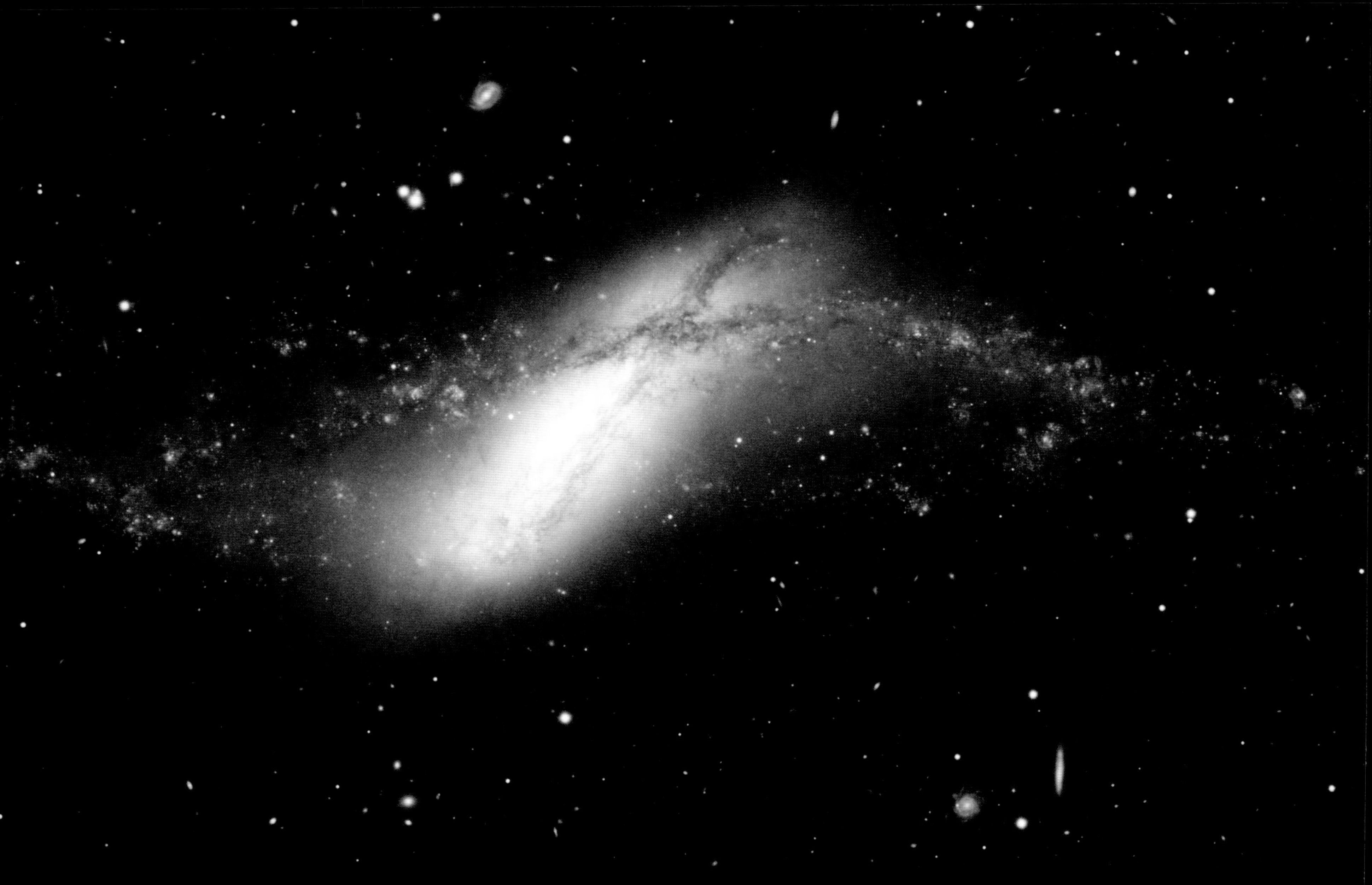

Galaxy Birth, Death, and Reincarnation

The births and deaths of galaxies are intertwined. Large galaxies grow larger by devouring smaller ones in brazen acts of galactic cannibalism. Shocks from collisions between passing galaxies ignite the birth of new stars that light up the night sky. Material stripped from galaxies accumulates over time, creating a sea of orphaned stars and gas in space. Yet the demise of some galaxies also gives rise to others, as debris is recycled into existing or new generations of galaxies.

The galaxy seen here, NGC 660, bears the scars of a violent collision that probably occurred about a billion years ago. The ring of stars, gas, and dust that encircles NGC 660 may be the shredded remains of a neighboring galaxy that strayed too close or perhaps a ripple splashed into space when two smaller galaxies merged to form the object we see today.

The Gemini North telescope snapped this half-hour exposure with its Gemini Multi-Object Spectrograph, one of the observatory's most requested instruments. IMAGE: GEMINI OBSERVATORY/AURA/TRAVIS RECTOR

Beyond the Rainbow

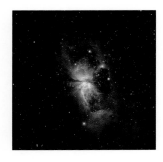

"It often seems to me that the night is even more richly colored than the day," wrote Vincent van Gogh, the great Dutch artist whose *Starry Night* is one of the world's most beloved paintings.

In fact, planets, stars, and galaxies come in an even wider range of colors than our eyes can see. Light takes many forms including gamma rays, X-rays, ultraviolet, infrared, submillimeter, and radio waves that are invisible to us but not to specially designed instruments on telescopes. Each type of light provides a different view of the cosmos.

The Subaru Telescope's CISCO camera captured this infrared image of a stellar nursery known as S106 IRS4, a region where stars are being born today from the ashes of earlier generations. These newborn stars warm the gas that surrounds them, causing it to glow in a kaleidoscope of colors more vibrant than van Gogh could ever have imagined. IMAGE: SUBARU TELESCOPE/NAOJ

Untwinkling the Stars

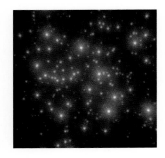

This image of the Arches Cluster, taken with the Keck II telescope, demonstrates the remarkable power of adaptive optics to sharpen our view of the universe.

After a journey lasting millions or billions of years through mostly empty space, light from distant stars and galaxies must plow through Earth's atmosphere in its final moments to reach telescopes on the ground. Swirling, turbulent pockets of air jostle this incoming light, making the stars appear to twinkle and blurring our view of the cosmos.

At a height of 13,796 feet (4,205 meters) above sea level, Maunakea rises above 40 percent of our planet's atmosphere, which allows telescopes there to reach up and collect light from the universe before the thicker layers of air below can distort it. The remaining atmosphere above Maunakea is also exceptionally stable, allowing sharper images to be obtained there than most other places in the world.

But it's possible to do even better. Many of the telescopes on Maunakea use a technique known as adaptive optics to eliminate much of the blurriness caused by the atmosphere, yielding views of the heavens as detailed as any seen from space.

The stars in this image, taken with Keck Observatory's adaptive optics system, are so sharp that they appear as tiny pinpoints of light, approaching the highest resolution that is theoretically achievable. The faint halos surrounding the stars are all that remains after Keck Observatory's adaptive optics system removed most of the distortions caused by our planet's atmosphere.

The Arches Cluster is the densest known star cluster in our Milky Way galaxy, home to about 150 young heavyweight stars as well as thousands of smaller ones. Located less than 100 light-years from the center of the Milky Way and the giant black hole that resides there, the Arches Cluster will eventually be torn apart by the intense gravity in this region, scattering its stars into space.
IMAGE: W. M. KECK OBSERVATORY AND MARK MORRIS

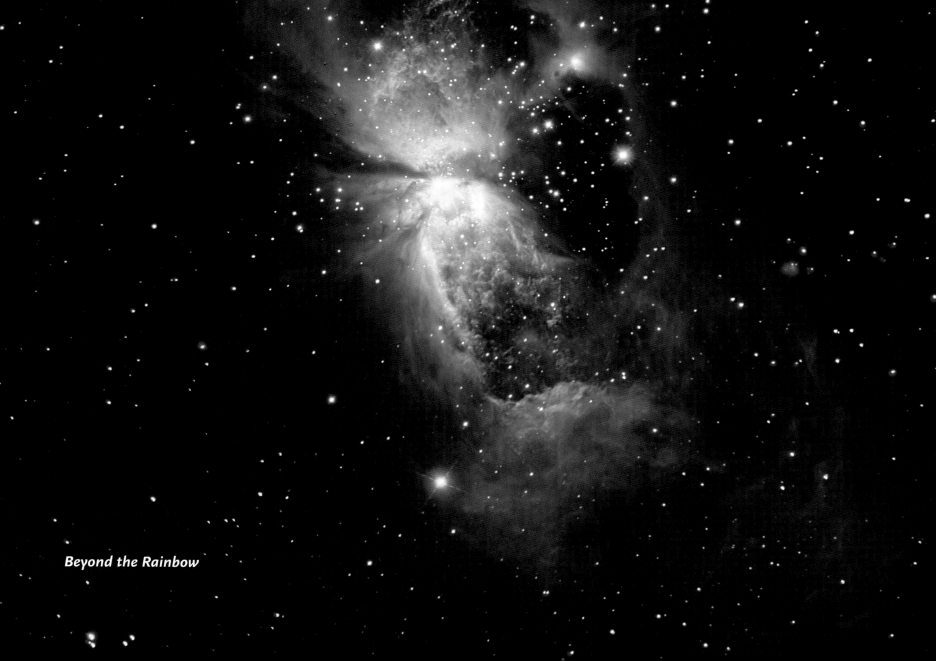

Beyond the Rainbow

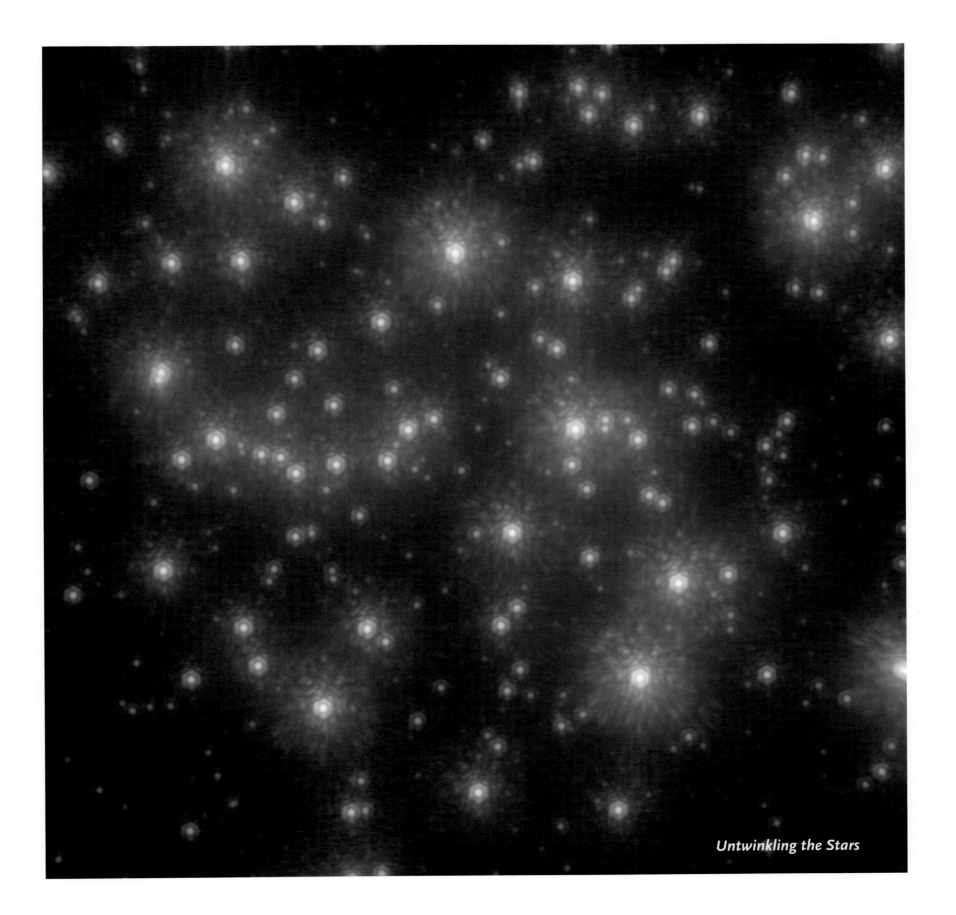

Untwinkling the Stars

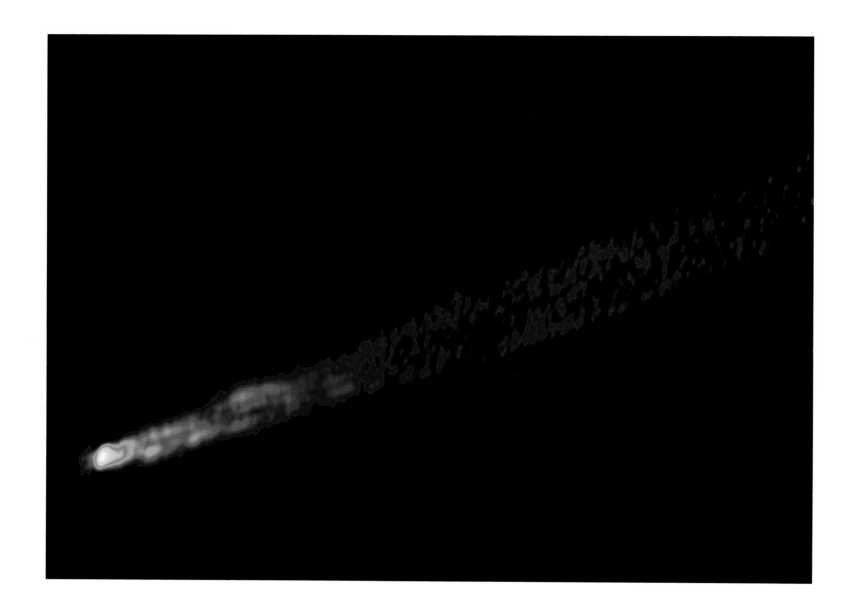

M87 is a giant galaxy about 50 million light-years away in the constellation Virgo. In 1918, astronomer Heber Curtis discovered a powerful jet of material coming from the galaxy's center, which he described as "a curious straight ray."

Today, we know that this jet is created by an enormous black hole weighing more than 6 billion suns. Material drawn to this region by the black hole's immense gravity forms a rapidly rotating disk that releases energy by blasting streams of subatomic particles at nearly the speed of light.

The Very Long Baseline Array (VLBA) combined light from its ten radio telescopes distributed across North America, including one on Maunakea, to capture this image of the innermost part of M87's jet. This technique, known as interferometry, provides extremely high resolution views of the cosmos, including this look into the mysterious regions surrounding a black hole. IMAGE: NRAO/AUI, Y. Y. KOVALEV, MPIFR, AND ASC LEBEDEV

A Cosmic Jet

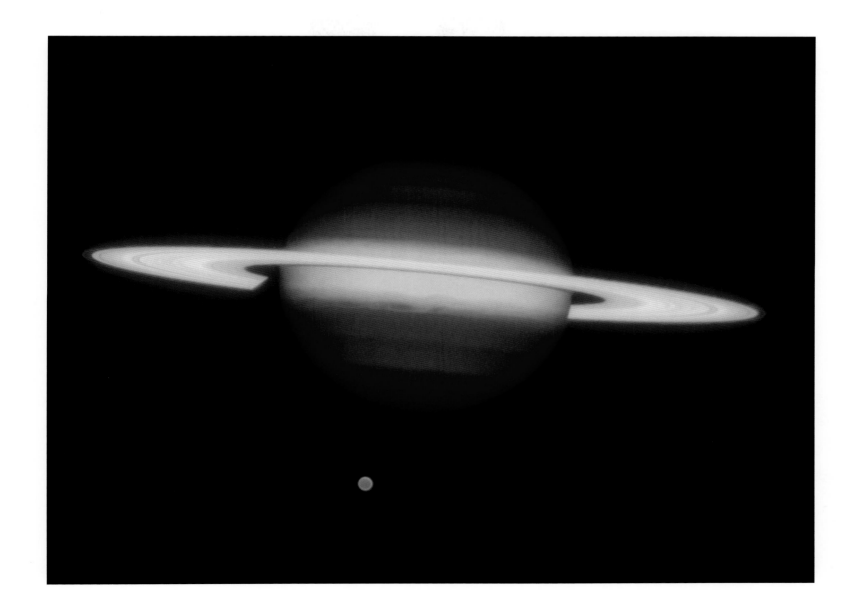

Saturn and its rings are among the most spectacular sights in the night sky. This infrared image, taken with the Gemini North telescope in 2009, shows the ringed planet accompanied by its largest moon, Titan. The only moon in our solar system with a substantial atmosphere, Titan is considered one of the most promising places to find life beyond Earth. IMAGE: GEMINI OBSERVATORY/AURA; HENRY ROE; LOWELL OBSERVATORY; EMILY SCHALLER; INSTITUTE FOR ASTRONOMY, UNIVERSITY OF HAWAI'I

Saturn and Titan

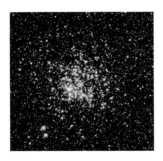

Jewel of the Night

M11 is a group of several thousand hot young stars located about 5,000 light-years from Earth. These siblings, born about 250 million years ago, were discovered by German astronomer Gottfried Kirch in 1681 and recorded by Charles Messier in 1764. M11 is one of the richest and most compact star clusters in the Milky Way. PHOTO: CANADA-FRANCE-HAWAI'I TELESCOPE/COELUM, J.-C. CUILLANDRE, AND G. ANSELMI

A Look Back in Time

A portion of the Subaru Deep Field—obtained by observing a patch of sky with the Subaru Telescope over a total of thirty nights between April 2002 and March 2004—reveals objects of all shapes, sizes, and colors strewn throughout space. Most of the objects in this image are galaxies, billions of light-years away, which we see in their infancy. The Subaru Deep Field has yielded many scientific discoveries, including the identification of some of the most distant known galaxies and the detection of 10-billion-year-old supernova explosions. PHOTO: SUBARU TELESCOPE/NAOJ AND R. JAY GABANY

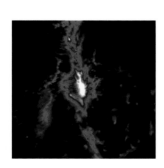

A River of Stars

"The stars go streaming through space pulsed on and on forever like blood globules in Nature's warm heart," wrote John Muir, nineteenth-century naturalist.

Stars are round in shape, but the cold clouds of gas and dust in which they form usually aren't. The Orion A region, seen here in submillimeter light captured by the SCUBA-2 instrument of the James Clerk Maxwell Telescope (JCMT), is an elongated region of ongoing star formation. Stars of all sizes are currently being born there, including some much larger than our sun. Over the past few decades, observations of star-forming regions like this have revealed that most embryonic stars are found in filamentary condensations of gas and dust. The origin of these filaments and their role in star formation are still open questions. PHOTO: JOINT ASTRONOMY CENTRE AND CARL SALJI ET AL.

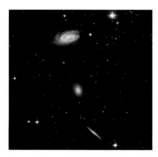

A Moment Frozen in Time

"Time itself comes in drops," wrote William James, the American philosopher and psychologist.

One of those drops of time was captured by the Canada-France-Hawai'i Telescope (CFHT) in this 2008 portrait of a group of galaxies in the constellation Draco. The light recorded by CFHT's MegaCam camera began its long journey to Earth almost 100 million years ago, showing these galaxies as they looked then, not now. It's a moment frozen in time, a snapshot of the history of our universe. PHOTO: CANADA-FRANCE-HAWAI'I TELESCOPE/COELUM, J.-C. CUILLANDRE, AND G. ANSELMI

Jewel of the Night

A Look Back in Time

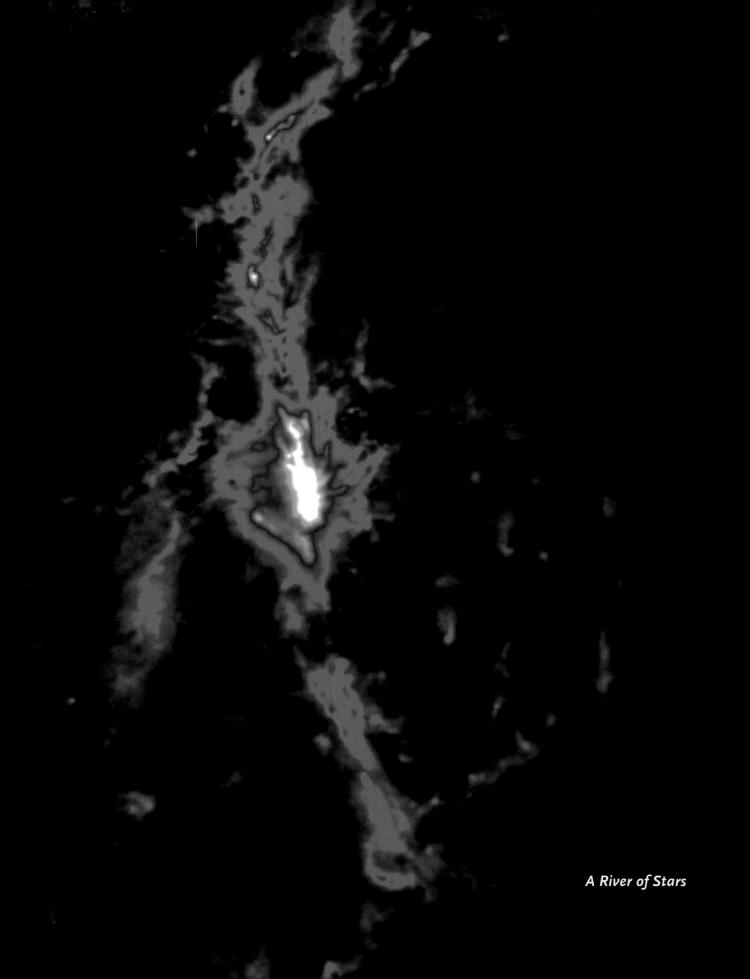

A River of Stars

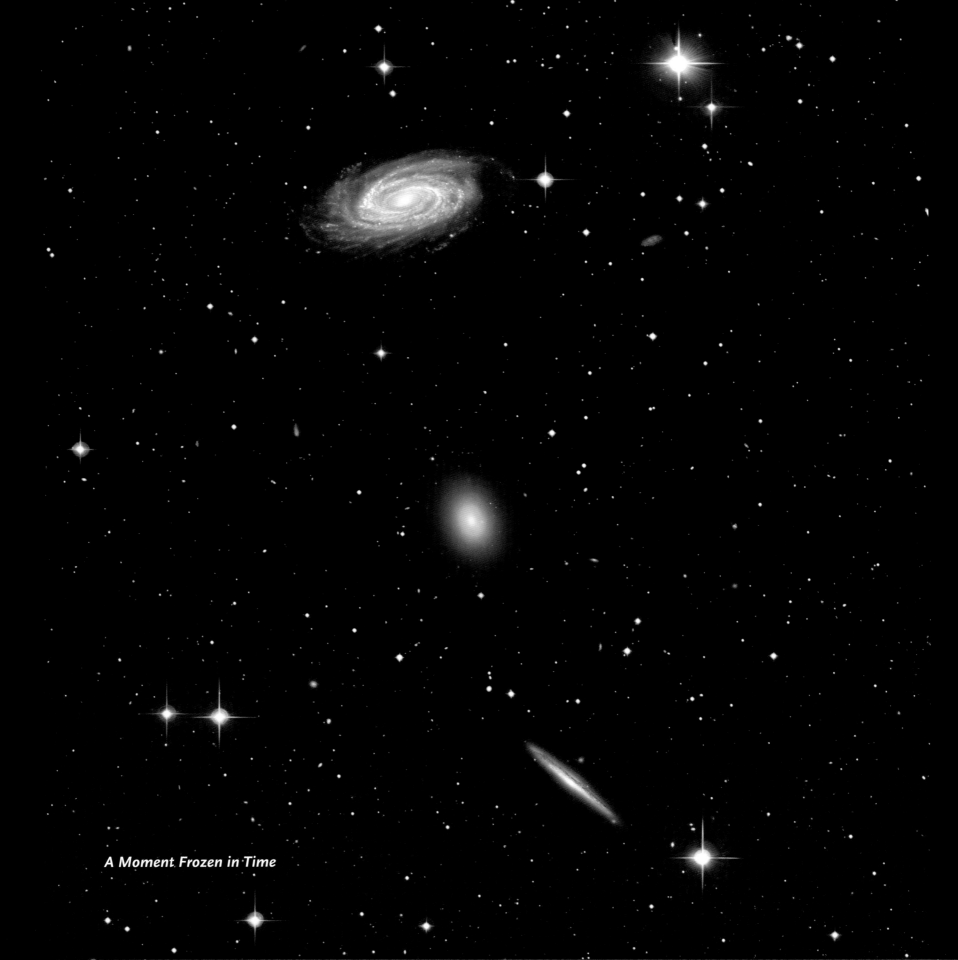

A Moment Frozen in Time

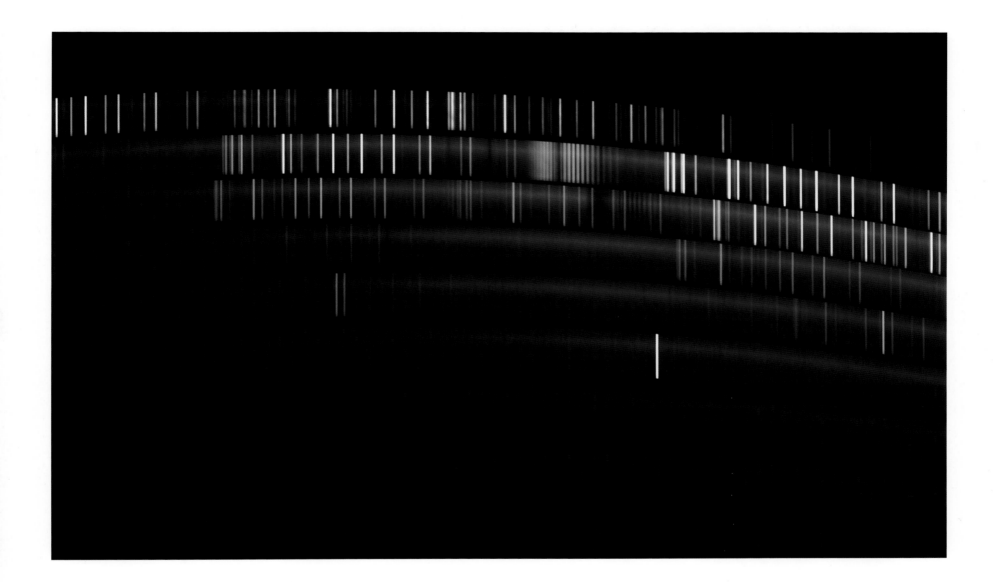

Legend has it that when he took his first sip of champagne, the seventeenth-century French monk Dom Pérignon exclaimed, "Come quickly, I am tasting the stars!"

Unfortunately, astronomers can't taste or touch the stars or most other objects in the night sky. Instead, information about them must be extracted from the light they emit. A star's light, for example, reveals how far away it is, what it's made of, how hot it is, how fast it's moving through space, and even how old it is.

This image shows the light from a galaxy as seen with the Keck II telescope. A special instrument called the Echellette Spectrograph and Imager acts like a high-tech prism to disperse the light into its component colors, providing astronomers with a wealth of information about the object. PHOTO: MICHAEL WEST

Decoding Starlight

Strange Symmetry

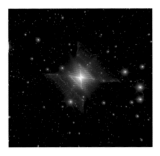

Looking back over his illustrious career at age seventy, the great Dutch astronomer Jan Oort said, "One thing I've learned in my life is that one must be prepared to meet entirely unexpected things."

Images of a star known as MWC 922, taken with the Keck II telescope on Maunakea, were combined with observations from the Mount Palomar Hale telescope in California to reveal this exotic object known as the "Red Square Nebula." Located about 5,000 light-years from Earth, the distinctive appearance of this nebula—the Latin word for "cloud"—is created as the star illuminates gaseous material flowing out from it.

Exquisitely sharp images like this are possible thanks to adaptive optics, a technique that reduces blurriness introduced by our planet's atmosphere whenever light passes through it. Adaptive optics systems on Keck Observatory and other Maunakea telescopes routinely produce images as sharp or sharper than those obtainable with the Hubble Space Telescope orbiting high above our planet's atmosphere. PHOTO: PETER TUTHILL, SYDNEY UNIVERSITY PHYSICS DEPARTMENT, PALOMAR OBSERVATORY, AND W. M. KECK OBSERVATORY

Seeing with Submillimeter Eyes

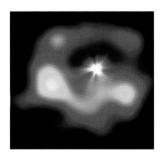

This image, taken in submillimeter light by the James Clerk Maxwell Telescope (JCMT), shows a ring of dust particles around a nearby sunlike star named Epsilon Eridani. The clumpy structure in the ring is suggestive of planets forming around this star, and it is already known from other observations that there is at least one planet orbiting Epsilon Eridani. The bright object at the center of the ring is an artist's rendition of the star. PHOTO: JOINT ASTRONOMY CENTRE AND JANE GREAVES ET AL.

A Plethora of Planets

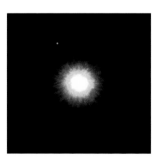

When the first telescopes went into operation on Maunakea, no planets were known beyond our solar system. Today, thanks to pioneering work by telescopes there and elsewhere, thousands of planets have been found circling other stars.

But relatively few of those planets have been photographed directly. Most are too faint to be seen against the backdrop of their parent star's intense glare; their existence can only be inferred indirectly. New techniques, however, allow astronomers to block much of a star's brilliance, revealing planets huddled around them.

In 2008, the Gemini North telescope captured this image of a planet orbiting a nearby sunlike star using the Near-Infrared Imager (NIRI) camera with adaptive optics to remove blurriness that occurs when light passes through Earth's atmosphere. This planet, whose mass is about eight times that of Jupiter, orbits much farther away from its star than any of the sun's planets. PHOTO: GEMINI OBSERVATORY/AURA, D. LAFRENIERE, R. JAYAWARDHANA, AND M. VAN KERKWIJK

Strange Symmetry

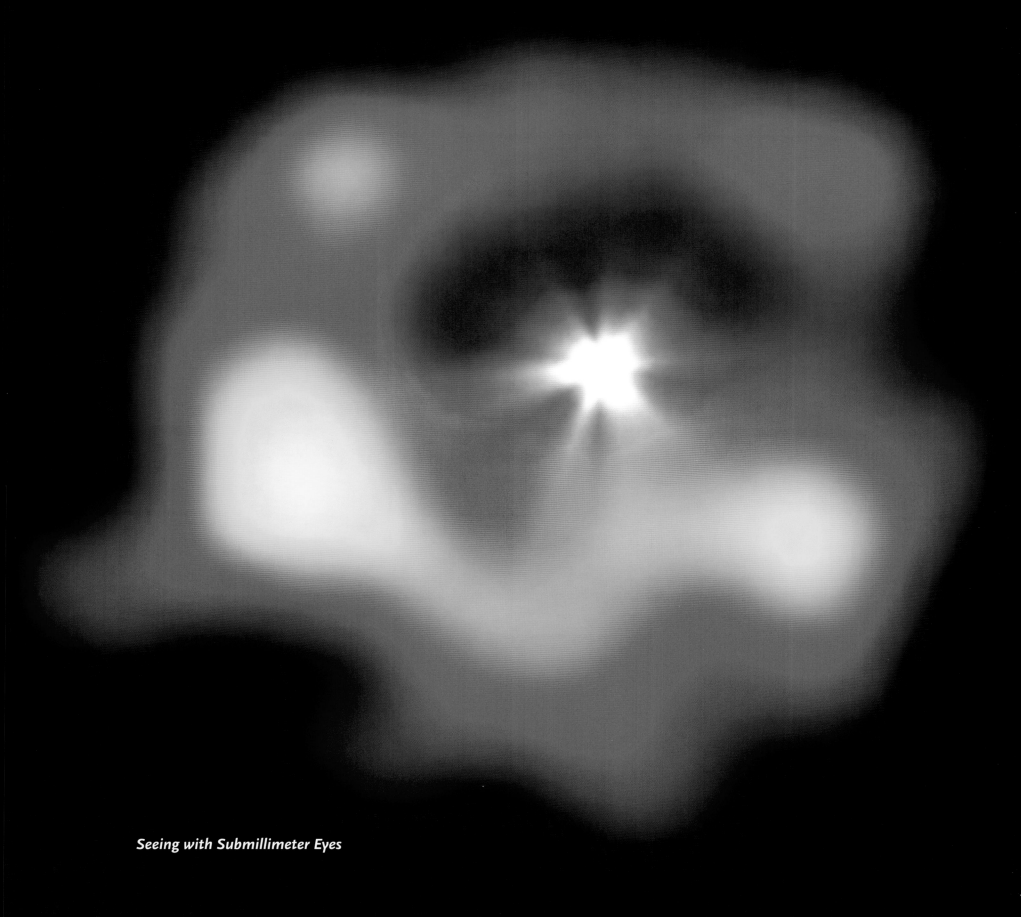

Seeing with Submillimeter Eyes

A Plethora of Planets

Revealing the Hidden Universe

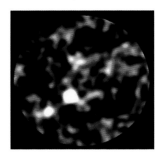

Nothing travels faster than light—a light beam could circle the earth more than seven times in one second—but the universe is so immense that light from distant objects still takes millions or billions of years to reach us. Everything we see, we see as it appeared in the past. We see the moon as it looked a second ago and the sun as it appeared eight minutes ago. And the farther we look into space, the more we look back in time.

In 1995, the Hubble Space Telescope took the deepest and most detailed snapshot yet of the young universe. Known as the Hubble Deep Field, this revolutionary image revealed thousands of distant and faint galaxies in a tiny sliver of sky. But even Hubble's sharp eye can't see everything. The visible light of very distant galaxies is shifted into the submillimeter part of the spectrum as the universe expands. Many newborn galaxies are also hidden behind vast clouds of gas and dust that makes them impossible to see in the visible light captured by Hubble. To see the distant, dusty universe requires collecting other types of light, such as infrared, submillimeter, and radio waves.

This image shows the Hubble Deep Field as seen by the SCUBA instrument on the James Clerk Maxwell Telescope (JCMT), which detects submillimeter light that's able to penetrate the dust surrounding young galaxies, allowing astronomers to glimpse their birth and growth. PHOTO: JOINT ASTRONOMY CENTRE AND DAVID HUGHES ET AL.

Comet ISON's Swan Song

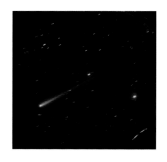

According to Greek legend, when Icarus and his father Daedalus were imprisoned on the island of Crete, Daedalus built wings of feathers and wax for their escape, cautioning Icarus not to fly too high because the sun would melt the wax. But Icarus was so overjoyed by his ability to soar and swoop like a bird that he forgot his father's warning. As he flew higher and higher, the feathers came loose and he fell to his death in the sea below.

Like Icarus, Comet ISON flew too close to the sun and perished. The mile-wide chunk of ice was once a prisoner too, held for billions of years in our solar system's dark outer regions. Freed by a gravitational tug from a passing star, the comet began its exhilarating but ill-fated flight a few million years ago. As it passed near the sun in late 2013, the intense heat vaporized the comet, its atoms exhaled into space like a dying breath. Ice and dust, like feathers and wax, proved no match for infernal heat. Predicted to become one of the brightest comets in recent memory, the comet never lived up to expectations, illuminating the darkness only briefly before dissolving.

This stunning portrait of Comet ISON was captured by the Subaru Telescope's wide-field imaging camera Hyper Suprime-Cam mounted at the telescope's prime focus. Stars appear as streaks rather than points of light in this image because the telescope was set to track the moving comet. PHOTO: HSC PROJECT AND SUBARU TELESCOPE/NAOJ

Stormy Jupiter

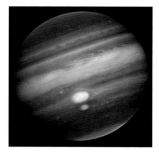

Jupiter, the largest planet in our solar system, is a mass of swirling gas. More than 1,300 Earth's could fit inside it. This razor-sharp infrared image, a color composite obtained by the Gemini North telescope, reveals a wealth of details about Jupiter, including bands of clouds and two large storms in the giant planet's atmosphere. PHOTO: GEMINI OBSERVATORY/AURA

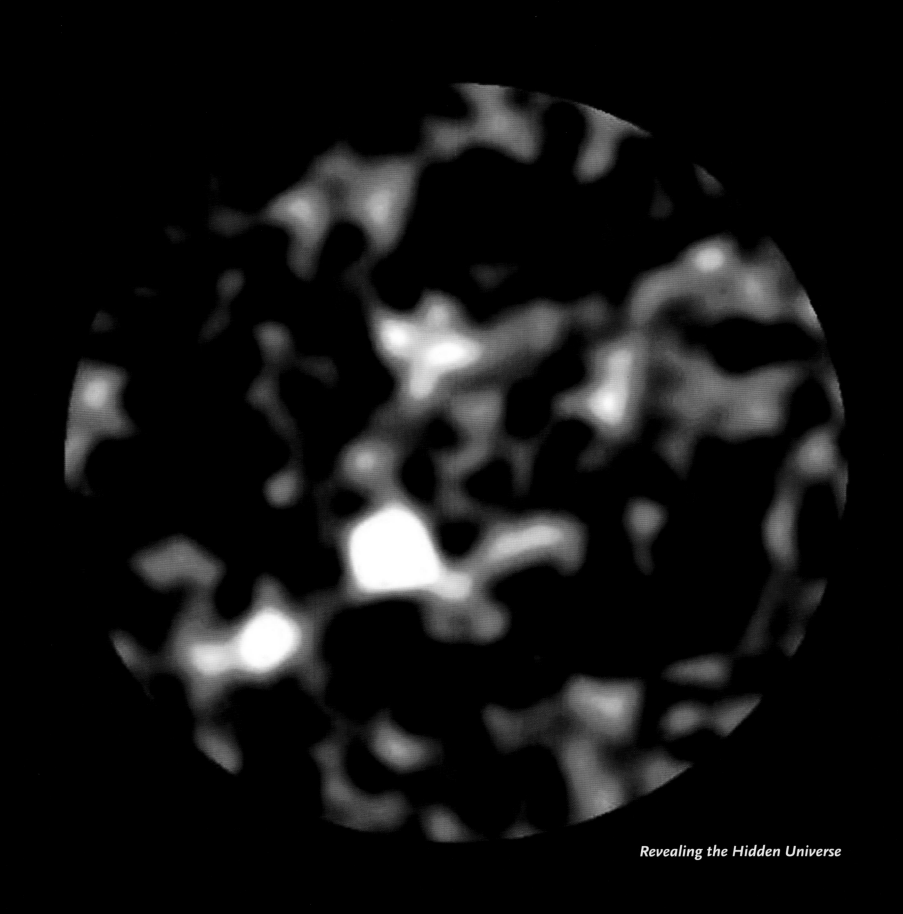

Revealing the Hidden Universe

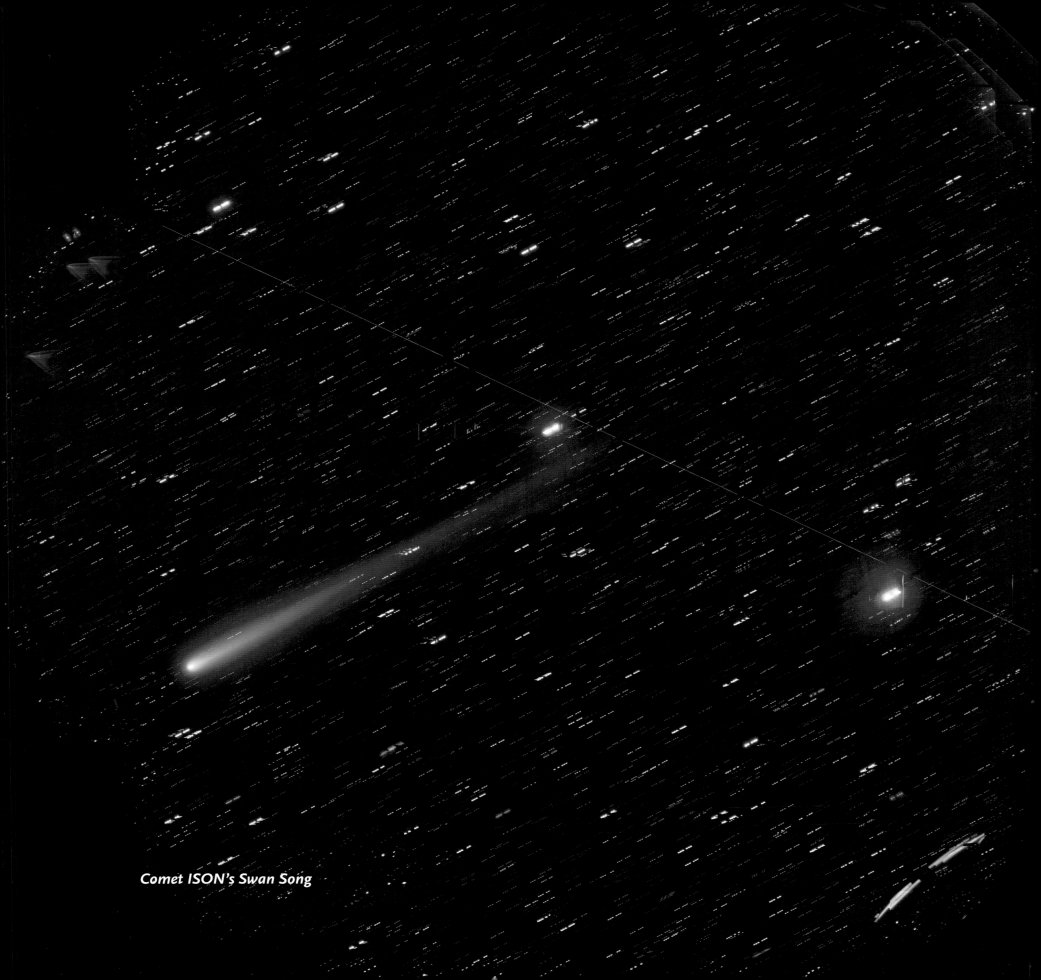

Comet ISON's Swan Song

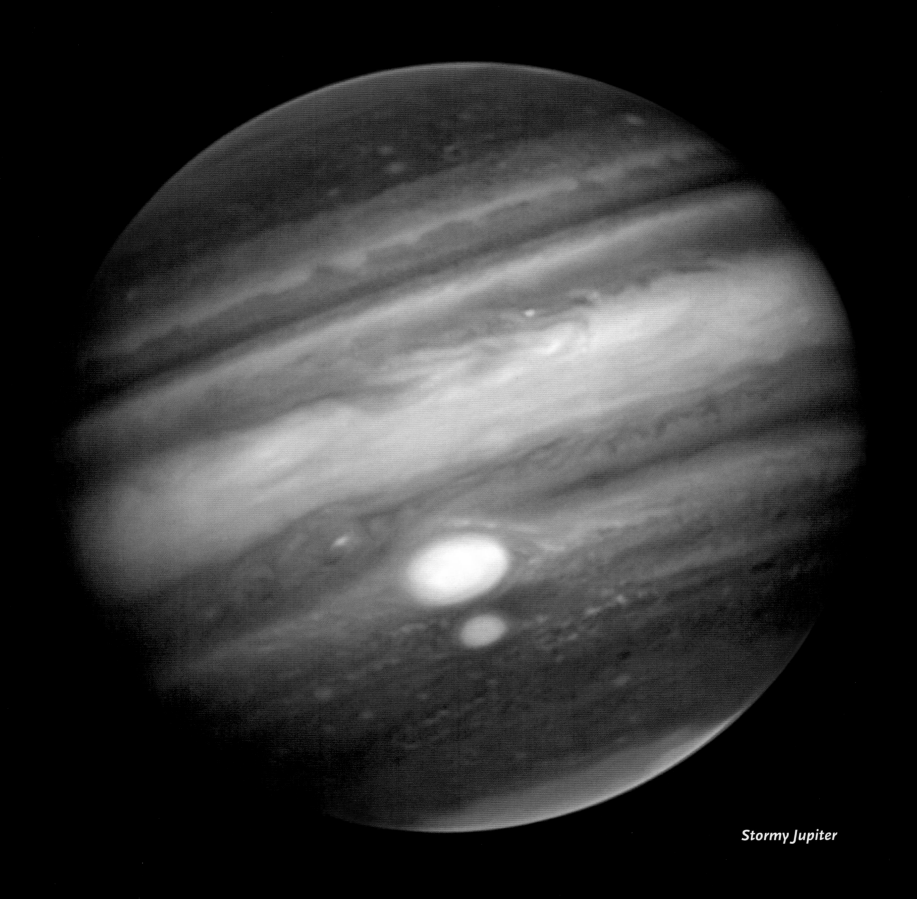

Stormy Jupiter

The Ghosts
of Galaxies

"Nothing comes out more clearly in astronomical observations than the immense activity of the universe," said Maria Mitchell, the first woman astronomer in the United States.

This image of the nearby spiral galaxy M81, taken by the Subaru Telescope in 2005, provides a good example of such activity. The diffuse glow and luminous arcs of material seen in M81's outskirts are probably the ghostly remains of smaller galaxies that were cannibalized in the past, or perhaps the aftermath of an interaction between M81 and its neighbor M82 that occurred millions of years ago. PHOTO: SUBARU TELESCOPE/NAOJ

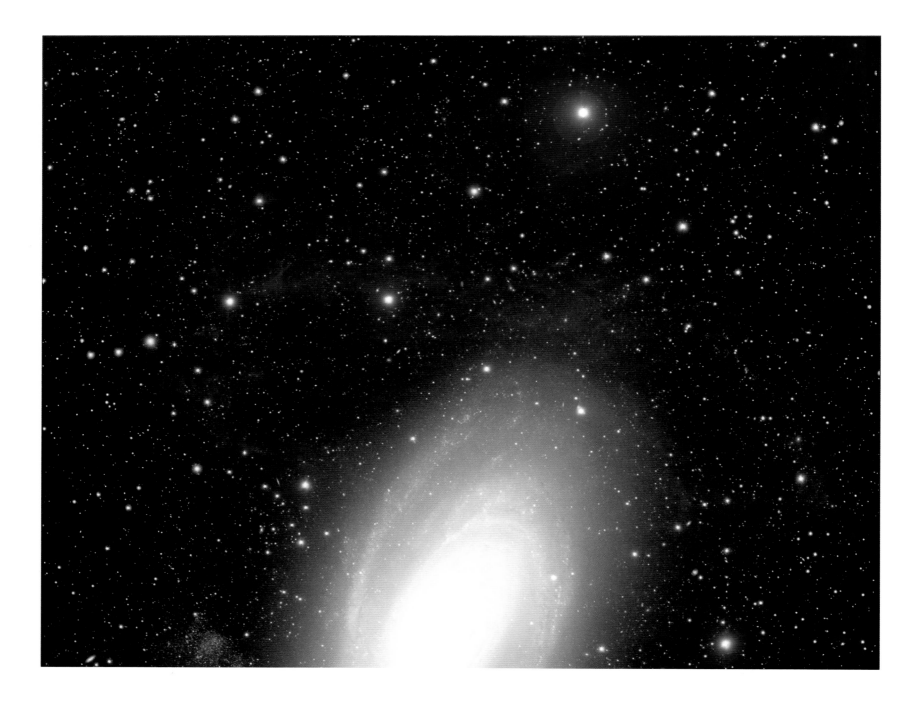

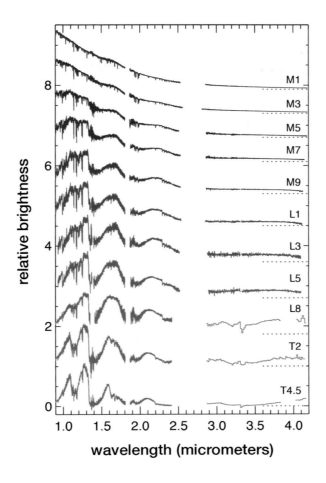

Light is composed of waves, and different kinds of light have different wavelengths. For example, X-rays, ultraviolet light, and visible light have shorter wavelengths than infrared, submillimeter, and radio waves.

Spectroscopy—the detailed analysis of light—is one of the most important tools in astronomy, and telescopes on Maunakea spend a large fraction of their time making spectroscopic observations. An instrument called a spectrograph spreads incoming light into its component wavelengths, called a spectrum, which can be studied in great detail to reveal a wealth of information about the observed object, such as its temperature, chemical composition, speed, and other physical properties.

This image shows infrared spectra for objects known as M, L, and T dwarfs obtained with the SpeX instrument on the NASA Infrared Telescope Facility (IRTF). The curves labeled M1 through M9 are spectra of stars, showing how much light they emit at each different wavelength. Spectra labeled L1 through T4.5 are from brown dwarfs, objects with less mass than stars but more mass than planets. The differences in the spectra seen here reflect changes in the temperature and atmospheric chemistry of these objects. The faintest brown dwarfs are only about seven times more massive than Jupiter and have much cooler temperatures than stars at only about 80 degrees Fahrenheit (27 degrees Celsius)—room temperature! IMAGE: IRTF AND MICHAEL CUSHING

More than Just Pretty Pictures

The Light of
Dead Stars

Every star dies sooner or later. After shining for millions or billions of years, they eventually run out of fuel. Some stars end their lives with violent explosions, flaring briefly but brilliantly. Most, however, simply fade away. Yet the light emitted during a star's lifetime continues its journey through space long after the star's internal flame has gone out, illuminating the sky like a memory, flickering but not forgotten.

This image, taken with the Canada-France-Hawai'i Telescope (CFHT), shows an ancient cluster of stars known as NGC 6144. Located 28,000 light-years from Earth, NGC 6144 is seen through gas that was shed by the nearby star Antares, coloring the night in that region of the sky. IMAGE: CANADA-FRANCE-HAWAI'I TELESCOPE/COELUM, J.-C. CUILLANDRE, AND G. ANSELMI

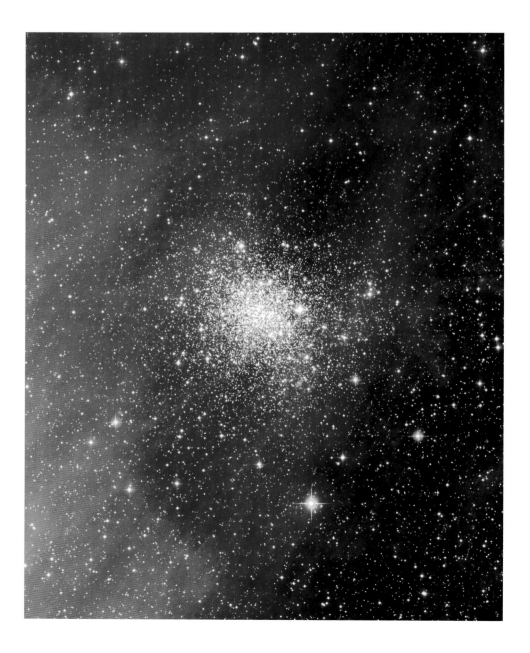

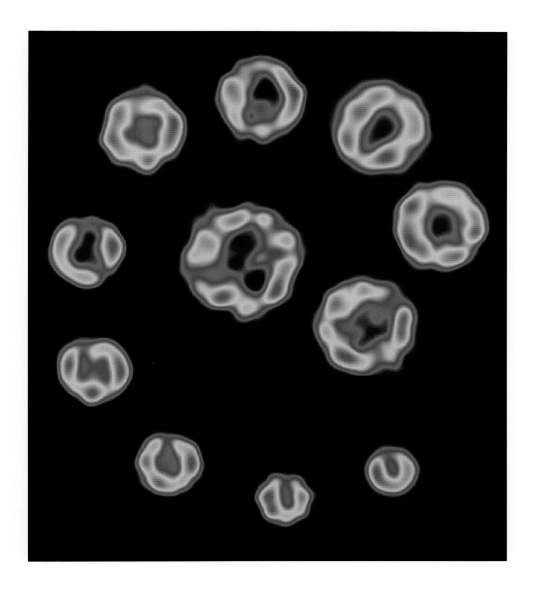

Supernova 1993J, the explosive death of a massive star in the nearby galaxy M81, was discovered in March 1993 by an amateur astronomer in Spain. Thanks to the tremendous resolving power of the Very Long Baseline Array (VLBA), this network of ten radio telescopes—including one on Maunakea—was able to study radio emission from the blast as it expanded at thousands of miles per second over the next seven years.

This VLBA image shows a time sequence of the supernova's growing sphere of shrapnel, beginning with a 1993 image at the bottom right and following a spiral pattern that culminates with the image at the center taken in 2000. IMAGE: NRAO/AUI, N. BARTEL, M. BIETENHOLZ, AND M. RUPEN, ET AL.

Snapshots of
an Explosion

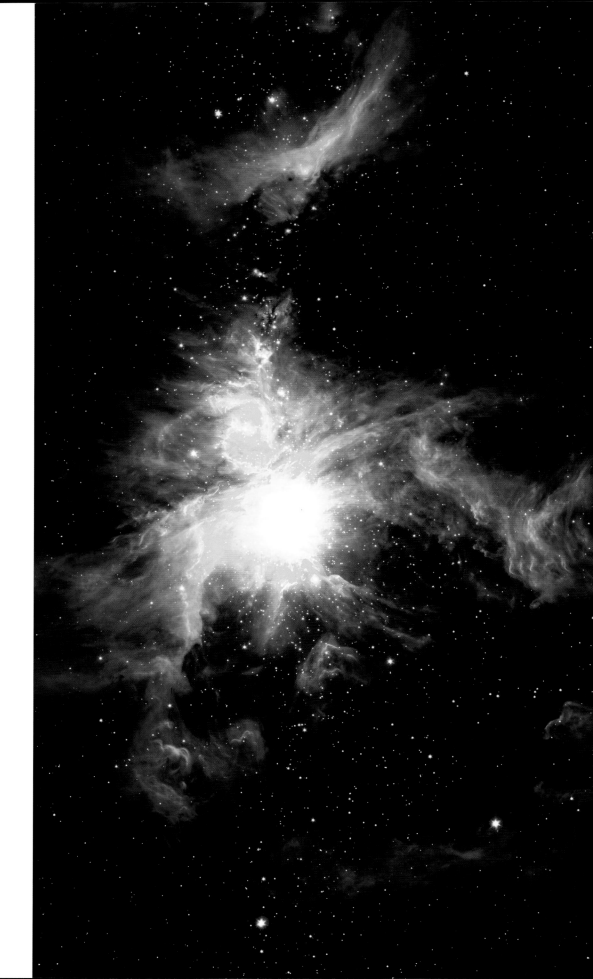

Seeing with New Eyes

"The voyage of discovery lies not in seeking new horizons, but in seeing with new eyes," said Marcel Proust, the great French novelist.

This stunning image shows a portion of the Orion Nebula, a region of intense star formation located about 1,300 light-years from Earth. The image is a composite of light from three different telescopes: the United Kingdom Infrared Telescope (UKIRT) on Maunakea, the IRAM Millimeter-wave Telescope in Spain, and the orbiting Spitzer Space Telescope. Each collects light that our eyes can't see but that young stars emit in abundance. IMAGE: CHRIS DAVIS AND JOINT ASTRONOMY CENTRE/IRAM/SPITZER

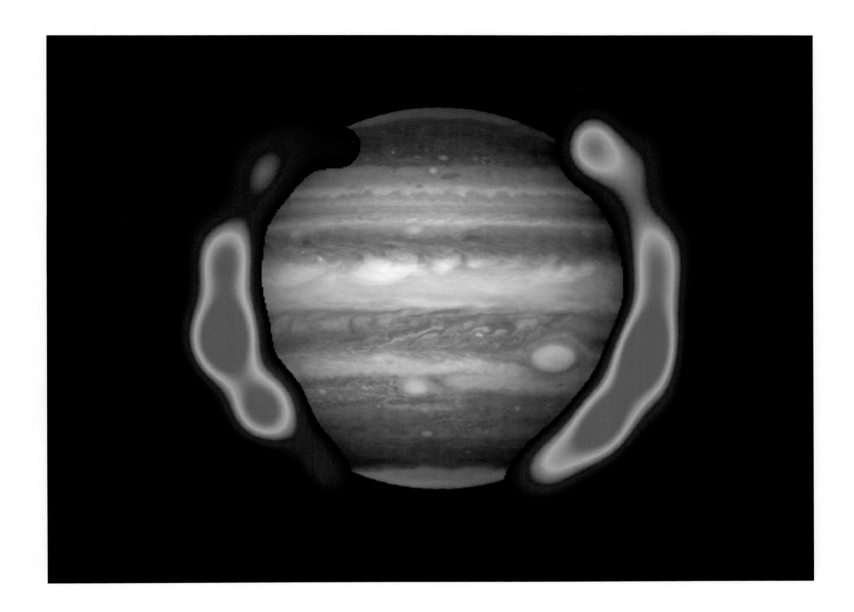

When fragments of Comet Shoemaker–Levy 9 slammed into Jupiter in July 1994, they deposited an amount of energy equivalent to millions of megatons of TNT, churning the giant planet's atmosphere and sending fireballs of material sky high.

Fifteen years later, the giant planet was still showing the effects of the encounter. This 2009 composite image shows Jupiter as seen in visible light by the Hubble Space Telescope, with submillimeter light captured by the Submillimeter Array (SMA) superimposed. The red, blue, and yellow colors around Jupiter's periphery show the distribution of hydrogen cyanide molecules high in the planet's atmosphere. These molecules are good tracers of dense gas and are frequently used to map star-forming regions in galaxies. Although hydrogen cyanide is plentiful around most of the planet, there's a noticeable lack near the north and south poles, suggesting the possible presence of a polar vortex. IMAGE: SUBMILLIMETER ARRAY AND MARK GURWELL

Jupiter's Afterglow

INDEX

ABOUT THE AUTHOR

MICHAEL J. WEST is the Deputy Director for Science at Lowell Observatory in Flagstaff, Arizona. He obtained his PhD in astronomy from Yale University in 1987 and has held teaching and research positions around the world, including seven years as a professor at the University of Hawai'i. A frequent user of the telescopes on Maunakea, he is the author of a previous work, *A Gentle Rain of Starlight: The Story of Astronomy on Mauna Kea* (2005).

Production Notes for
West / A SKY WONDERFUL WITH STARS

Book design and composition by Mardee Melton with display type
in Scala Sans Pro and body text in 10-point Bembo MT Pro.

Printing and binding by Regent Publishing Services
Printed on 140gsm Gold East Matte